Advances in MATHEMATICAL ECONOMICS

Aims and Scope. The project is to publish *Advances in Mathematical Economics* once a year under the auspices of the Research Center of Mathematical Economics. It is designed to bring together those mathematicians who are seriously interested in obtaining new challenging stimuli from economic theories and those economists who are seeking effective mathematical tools for their research.

The scope of *Advances in Mathematical Economics* includes, but is not limited to, the following fields:

- Economic theories in various fields based on rigorous mathematical reasoning.
- Mathematical methods (e.g., analysis, algebra, geometry, probability) motivated by economic theories.
- Mathematical results of potential relevance to economic theory.
- Historical study of mathematical economics.

Authors are asked to develop their original results as fully as possible and also to give a clear-cut expository overview of the problem under discussion. Consequently, we will also invite articles which might be considered too long for publication in journals.

Springer Japan KK

S. Kusuoka, T. Maruyama (Eds.)

Advances in Mathematical Economics

Volume 5

Springer

Shigeo Kusuoka
Professor
Graduate School of Mathematical Sciences
University of Tokyo
3-8-1 Komaba, Meguro-ku
Tokyo, 153-0041 Japan

Toru Maruyama
Professor
Department of Economics
Keio University
2-15-45 Mita, Minato-ku
Tokyo, 108-8345 Japan

ISBN 978-4-431-67976-9   ISBN 978-4-431-53979-7 (eBook)

DOI 10.1007/978-4-431-53979-7

Printed on acid-free paper

Originally published by Springer-Verlag Tokyo Berlin Heidelberg New York in 2003
Softcover reprint of the hardcover 1st edition2003

Camera-ready copy prepared from the authors' LaTeXfiles.

SPIN: 10898895

Table of Contents

Research Articles

G. Carlier
Duality and existence for a class of mass transportation problems and economic applications 1

C. Castaing, A. G. Ibrahim
Functional evolution equations governed by m-accretive operators 23

T. Fujimoto, J. A. Silva, A. Villar
Nonlinear generalizations of theorems on inverse–positive matrices 55

L. Hurwicz, M. K. Richter
Implicit functions and diffeomorphisms without C^1 65

L. Hurwicz, M. K. Richter
Optimization and Lagrange multipliers: non-C^1 constraints and "minimal" constraint qualifications 97

S. Kusuoka
Monte Carlo method for pricing of Bermuda type derivatives 153

Historical Perspective

I. Mutoh
Mathematical economics in Vienna between the Wars 167

Subject Index 197

Adv. Math. Econ. 5, 1–21 (2003)

Advances in
MATHEMATICAL
ECONOMICS

Duality and existence for a class of mass transportation problems and economic applications

Guillaume Carlier

Université Bordeaux 1, MAB, UMR CNRS 5466 and Université Bordeaux IV, GRAPE, UMR CNRS 5113, Avenue Léon Duguit, 33608, Pessac, FRANCE
(e-mail: Guillaume.Carlier@math.u-bordeaux.fr)

Received: April 15, 2002
Revised: May 20, 2002

JEL classification: C61, C82

Mathematics Subject Classification (2000): 90C08, 90C46, 91B40

Abstract We establish duality, existence and uniqueness results for a class of mass transportations problems. We extend a technique of W. Gangbo [9] using the Euler Equation of the dual problem. This is done by introducing the *h-Fenchel Transform* and using its basic properties. The cost functions we consider satisfy a generalization of the so-called *Spence-Mirrlees condition* which is well-known by economists in dimension 1. We therefore end this article by a somehow unexpected application to the *economic theory of incentives*.

Key words: mass transportation, duality, general Fenchel transform, economic theory of incentives, Spence-Mirrlees condition

1. Introduction and main statement

1.1 Assumptions and notations

Let us first recall that, given a probability space $(\Omega_1, \mathcal{A}_1, \mu_1)$, a measurable space $(\Omega_2, \mathcal{A}_2)$ and a measurable map $f : \Omega_1 \to \Omega_2$, the push-forward of μ_1 through f, denoted $f\sharp\mu_1$ is the probability measure on $(\Omega_2, \mathcal{A}_2)$ defined by:

$$f\sharp\mu_1(B) := \mu_1(f^{-1}(B))$$

for every $B \in \mathcal{A}_2$.

In all the following, Ω is some bounded connected open subset of $\mathbb{R}^n$, and μ is some probability measure in Ω which is absolutely continuous with respect to the n-dimensional Lebesgue measure, with a positive Radon-Nikodym derivative with respect to the n-dimensional Lebesgue measure and such that $\mu(\partial\Omega) = 0$.

We are also given a compact Polish space Y, a Radon probability measure ν on Y and a function $h : \overline{\Omega} \times Y \to \mathbb{R}$ which satisfies:

$$h \in C^0(\overline{\Omega} \times Y, \mathbb{R}), \tag{1}$$

for every $\omega \subset\subset \Omega$ there exists $c(\omega) > 0$ such that for all $(x_1, x_2) \in \omega^2$

$$\sup_{y \in Y} |h(x_1, y) - h(x_2, y)| \leq c(\omega)\|x_1 - x_2\|, \tag{2}$$

for all $y \in Y$, $h(., y)$ is differentiable in Ω and for all $(y_1, y_2, x) \in Y^2 \times \Omega$

$$\frac{\partial h}{\partial x}(x, y_1) = \frac{\partial h}{\partial x}(x, y_2) \Rightarrow y_1 = y_2. \tag{3}$$

Assumption (3) plays an important role in the proofs and we shall see that it may be interpreted as a generalization of the well-known one of Spence and Mirrlees, this assumption was first introduced by Levin in [13].

Our aim is to study the following Monge's mass transportation problem:

$$(\mathcal{M}) \quad \sup_{s \in \Delta(\mu,\nu)} J(s) := \int_\Omega h(x, s(x))d\mu(x)$$

with:

$$\Delta(\mu, \nu) := \{s \text{ is a Borel map} : \Omega \to Y \text{ s.t. } s\sharp\mu = \nu\}.$$

The associated Monge-Kantorovich problem is the linear (relaxation of $(\mathcal{M})$) program:

$$(\mathcal{MK}) \quad \sup_{\gamma \in \Gamma(\mu,\nu)} K(\gamma) := \int_{\Omega \times Y} h(x, y)d\gamma(x, y)$$

with:

$$\Gamma(\mu, \nu) := \{\gamma \text{ is a Borel probabiliy measure on } \Omega \times Y \text{ s.t. } \pi_1\sharp\gamma = \mu,\ \pi_2\sharp\gamma = \nu\}$$

where $\pi_1(x, y) = x$, $\pi_2(x, y) = y$ for all $(x, y) \in \Omega \times Y$.

Finally, we define the (dual of $(\mathcal{M})$) problem:

$$(\mathcal{D}) \quad \inf_{(\psi,\phi) \in E_h} L(\psi, \phi) := \int_\Omega \psi d\mu + \int_Y \phi d\nu$$

with:

$$E_h := \{(\psi, \phi), \text{ real-valued measurable s.t. } \psi(x) + \phi(y) \geq h(x, y),\ \forall (x, y) \in \Omega \times Y\}.$$

1.2 Main result

If ψ is a given real-valued function defined on Ω, we define the h-Fenchel Transform of ψ, ψ^h by:

$$\psi^h(y) := \sup_{x\in\Omega} h(x,y) - \psi(x), \text{ for all } y \in Y.$$

In a similar way, if ϕ is a given real-valued function defined on Y, we define the $\check{h}$-Fenchel Transform of ϕ, $\phi^{\check{h}}$ by:

$$\phi^{\check{h}}(x) := \sup_{y\in Y} h(x,y) - \phi(y), \text{ for all } x \in \Omega.$$

Our main result can then be stated as follows:

Theorem 1 *Under assumptions (1), (2), (3) the following assertions hold:*
1) problems $(\mathcal{M})$, $(\mathcal{MK})$ and $(\mathcal{D})$ admit at least one solution,
2) $(\mathcal{D})$ is dual to $(\mathcal{M})$ and $(\mathcal{MK})$ in the sense:

$$\inf(\mathcal{D}) = \sup(\mathcal{M}) = \sup(\mathcal{MK}),$$

3) the minimum in $(\mathcal{D})$ is attained by a pair $(\overline{\psi}, \overline{\phi})$ such that:

$$\overline{\psi} = \overline{\phi}^{\check{h}}, \overline{\phi} = \overline{\psi}^{h}$$

there exists moreover some Borel map $\overline{s}$ from Ω to Y which satisfies:

$$\overline{\psi}(x) + \overline{\phi}(\overline{s}(x)) = h(x, \overline{s}(x)), \textit{for all } x \in \Omega,$$

$\overline{s} \in \Delta(\mu,\nu)$ and is a solution of $(\mathcal{M})$, and $(id,\overline{s})\sharp\mu$ is a solution of $(\mathcal{MK})$,
4) uniqueness also holds: if s is a solution of $(\mathcal{M})$ then $s = \overline{s}$ μ-a.e., $(id,\overline{s})\sharp\mu$ is the unique solution of $(\mathcal{MK})$, and if (ψ,ϕ) is a solution of $(\mathcal{D})$ then $\psi - \overline{\psi}$ (respectively $\phi - \overline{\phi}$) is equal to some constant μ-a.e. (respectively ν-a.e.).

In Section 2, technical lemmas are established and basic properties of the h-Fenchel transform are proved. In Section 3, the main result is proved. Finally, in Section 4, we adress a question arising in the economic theory of incentives and show how assumption (3) can be interpreted as a natural generalization of the Spence-Mirrlees condition. In this framework, our main result enables to prove a general re-allocation principle.

The problem of optimal measure preserving maps $(\mathcal{M})$ has received a lot of attention since related questions naturally arise in fluid mechanics [2], differential geometry (see [16] for relation with a classical result of Aleksandrov

[1]), shape optimization [4] , functional analysis [11], [12], probability [19] and economics. In the case $Y \subset \mathbb{R}^n$ and $h(x,y) = x \cdot y$, the problem was solved by Brenier [3] who proved the important Polar Factorization Theorem and existence and uniqueness of an optimal map which is the gradient of some convex potential. This result was then extended by Mc Cann and Gangbo [10] for costs of the form $c(x-y)$ with c strictly convex. The result stated in Theorem 1, is very much in that spirit since it expresses existence and uniqueness of an optimal allocation map which is a measurable selection of the h-subdifferential of some h-convex potential. Similar characterization results were obtained by V. Levin [13] using a different approach based on cyclical monotonicity and the relaxed problem $(\mathcal{MK})$.

2. Technical preliminaries and h-Fenchel transform

In what follows ψ will always denote some function : $\Omega \to \mathbb{R} \cup \{+\infty\}$ and ϕ some function : $Y \to \mathbb{R} \cup \{+\infty\}$.

Definition 1 *1) ψ is h-convex if and only if there exists a nonempty subset A of $Y \times \mathbb{R}$ such that:*

$$\psi(x) = \sup_{(y,t)\in A} h(x,y) + t, \textit{ for all } x \in \Omega.$$

2) ϕ is $\check{h}$-convex if and only if there exists a nonempty subset B of $\Omega \times \mathbb{R}$ such that:

$$\phi(y) = \sup_{(x,t)\in B} h(x,y) + t, \textit{ for all } y \in Y.$$

Remark. If ψ is h-convex then either ψ is identically $+\infty$ or it is bounded. Note also that finite $\check{h}$-convex potentials are l.s.c, hence ν-measurable.

Definition 2 *1) The h-Fenchel Transform of ψ, ψ^h, is the $\check{h}$-convex function defined by:*

$$\psi^h(y) := \sup_{x\in\Omega} h(x,y) - \psi(x), \textit{ for all } y \in Y.$$

2) The $\check{h}$-Fenchel Transform of ϕ, $\phi^{\check{h}}$ is the h-convex function defined by:

$$\phi^{\check{h}}(x) := \sup_{y\in Y} h(x,y) - \phi(y), \textit{ for all } x \in \Omega.$$

Obviously, Young's inequalities hold:

$$\psi(x) + \psi^h(y) \geq h(x,y), \text{ for all } (x,y) \in \Omega \times Y \tag{4}$$

and:

$$\phi^{\check{h}}(x) + \phi(y) \geq h(x,y), \text{ for all } (x,y) \in \Omega \times Y. \tag{5}$$

Proposition 1

$$(\psi^h)^{\check{h}}(x) = \sup\{f(x) : f \leq \psi, \ f \text{ is } h\text{-convex }\}, \text{ for all } x \in \Omega,$$

$$(\phi^{\check{h}})^h(y) = \sup\{g(x) : g \leq \phi, \ f \text{ is } \check{h}\text{-convex }\}, \text{ for all } y \in Y.$$

It follows that ψ (respectively ϕ) is h-convex (respectively $\check{h}$-convex) if and only if $\psi = (\psi^h)^{\check{h}}$ (respectively $\phi = (\phi^{\check{h}})^h$).

Proof.

First $(\psi^h)^{\check{h}}$ is h-convex and Young's inequality yields $(\psi^h)^{\check{h}} \leq \psi$ so that, if we define:

$$V(x) := \sup\{f(x) : f \leq \psi, \ f \text{ is } h\text{-convex }\}, \text{ for all } x \in \Omega, \tag{6}$$

then :

$$(\psi^h)^{\check{h}} \leq V \leq \psi. \tag{7}$$

Since V is h-convex, there exists a nonempty subset A of $Y \times \mathbb{R}$ such that:

$$V(x) = \sup_{(y,t)\in A} h(x,y) + t, \text{ for all } x \in \Omega.$$

Let $(y_0, t_0) \in A$ and $\xi := h(.,y_0) + t_0$ we have:

$$\psi \geq \xi \Rightarrow (\psi^h)^{\check{h}} \geq (\xi^h)^{\check{h}} \tag{8}$$

of course $\xi \geq (\xi^h)^{\check{h}}$ and since $\xi^h(y_0) = -t_0$ then $(\xi^h)^{\check{h}}(x) \geq h(x,y_0) + t_0 = \xi(x)$ for all x so $(\xi^h)^{\check{h}} = \xi$, with (8) we get $(\psi^h)^{\check{h}} \geq \xi$ and since (y_0, t_0) is arbitrary in A taking the supremum yields $(\psi^h)^{\check{h}} \geq V$ so that $V = (\psi^h)^{\check{h}}$ using (7). The characterization of $(\phi^{\check{h}})^h$ is proved in the same way.

□

Definition 3 *1) Define, for all $x \in \Omega$:*

$$\partial^h \psi(x) := \{y \in Y : \psi(x') - \psi(x) \geq h(x',y) - h(x,y), \text{ for all } x' \in \Omega\}$$

$\partial^h \psi(x)$ is called the h-subdifferential of ψ at x, and ψ is h-subdifferentiable at x if and only if $\partial^h \psi(x) \neq \emptyset$.

2) Define, for all $y \in Y$:

$$\partial^{\check{h}} \phi(y) := \{x \in \overline{\Omega} : \phi(y') - \phi(y) \geq h(x,y') - h(x,y), \text{ for all } y' \in Y\}.$$

Note that $\partial^h\psi(x)$ and $\partial^{\check{h}}\phi(y)$ can also be defined by:

$$\partial^h\psi(x) = \{y \in Y : \psi(x) + \psi^h(y) = h(x,y)\}$$

and:

$$\partial^{\check{h}}\phi(y) = \{x \in \overline{\Omega} : \phi^{\check{h}}(x) + \phi(y) = h(x,y)\}.$$

In particular, if ψ is h-convex and $x \in \Omega$, then $y \in \partial^h\psi(x)$ if only if $x \in \partial^{\check{h}}\psi^h(y)$.

Proposition 2 *Let ψ be h-convex and finite, the following assertions hold true:*

1) For all $x \in \Omega$, $\partial^h\psi(x)$ is nonempty and compact, and the restriction of the set-valued map $\partial^h\psi$ to every closed subset of Ω has a closed graph,

2) $\psi \in W^{1,\infty}_{\mathrm{loc}}(\Omega)$, and : $esssup_\omega|\nabla\psi| \leq c(\omega)$ for all $\omega \subset\subset \Omega$ and $c(\omega)$ is given by (2),

3) if ψ is differentiable at $x \in \Omega$ and $y \in \partial^h\psi(x)$ then

$$\nabla\psi(x) = \frac{\partial h}{\partial x}(x,y),$$

4) there exists some Borel map s_ψ such that for almost every $x \in \Omega$, $\partial^h\psi(x) = \{s_\psi(x)\}$ and $s_\psi(x) \in \partial^h\psi(x)$ for all $x \in \Omega$.

Proof.

Let A be some nonempty subset of $Y \times \mathbb{R}$ such that:

$$\psi(x) = \sup_{(y,t)\in A} h(x,y) + t, \text{ for all } x \in \Omega.$$

1) Fix $x \in \Omega$ and let (y_n, t_n) be some sequence of A such that $h(x,y_n)+t_n \to \psi(x)$ as $n \to +\infty$. Up to a subsequence we may assume that y_n converges to some $\overline{y} \in Y$ so that $t_n \to \overline{t} := \psi(x) - h(x,\overline{y})$. Let us show now that $\overline{y} \in \partial^h\psi(x)$. Let $x' \in \Omega$ for all n, $\psi(x') \geq h(x',y_n) + t_n$ passing to the limit we get $\psi(x') \geq \psi(x) + h(x',\overline{y}) - h(x,\overline{y})$ i.e. $\overline{y} \in \partial^h\psi(x)$; $\partial^h\psi(x)$ is clearly compact since Y is and h is continuous. The fact that the restriction of $\partial^h\psi$ to every closed subset of Ω has a closed graph is straightforward.

2) Let $\omega \subset\subset \Omega$ and

$$c(\omega) := \sup_{(x_1,x_2,y)\in\omega^2\times Y,\ x_1\neq x_2} |h(x_1,y) - h(x_2,y)| \cdot \|x_1 - x_2\|^{-1} < +\infty.$$

Let $(x_1, x_2) \in \omega^2$ we have:

$$\psi(x_1) = \sup_{(y,t)\in A} h(x_1,y) + t = \sup_{(y,t)\in A} h(x_2,y) + t + h(x_1,y) - h(x_2,y)$$

$$\le \psi(x_2) + c(\omega)\|x_1 - x_2\|$$

finally, reversing order of x_1 and x_2 yields the desired result.

3) Let $x \in \Omega$ be a point of differentiabilty of ψ and $y \in \partial^h\psi(x)$, let $k \in \mathbb{R}^n$, and $t \neq 0$ be such that $[x - tk, x + tk] \subset \Omega$:

$$\psi(x + tk) - \psi(x) = t\nabla\psi(x) \cdot k + o(t) \ge h(x + tk, y) - h(x, y)$$

$$= t\frac{\partial h}{\partial x}(x, y) \cdot k + o(t)$$

dividing by $t > 0$ and letting $t \to 0^+$ in the previous yields

$$\nabla\psi(x) \cdot k \ge \frac{\partial h}{\partial x}(x, y) \cdot k$$

similarly the converse inequality also holds taking $t \to 0^-$ and since k is arbitrary we get:

$$\nabla\psi(x) = \frac{\partial h}{\partial x}(x, y).$$

4) By 2) and Rademacher's Theorem ψ is differentiable a.e. in Ω. On the other, hand since Y is compact and separable and using 1), $\partial^h\psi$ admits a measurable selection say s_ψ (see [6] or the measurable selection Theorem of Brown and Purves in [25]). If ψ is differentiable at $x \in \Omega$ and $y \in \partial^h\psi(x)$ then by 3) we get:

$$\frac{\partial h}{\partial x}(x, y) = \frac{\partial h}{\partial x}(x, s_\psi(x))$$

so that with (3) $\partial^h\psi(x) = \{s_\psi(x)\}$.

□

Corollary 1 *Let ψ_1 and ψ_2 be h-convex and finite, if for μ–a.e. $x \in \Omega$*

$$\partial^h\psi_1(x) \cap \partial^h\psi_2(x) \neq \emptyset$$

then $\psi_1 - \psi_2$ is constant.

Proof.

Using Proposition 2, we get $\nabla(\psi_1 - \psi_2) = 0$ a.e. in Ω hence the desired result, since Ω is connected.

□

We end this section by a result which will play a crucial role in the proof of the main result. The next Proposition is actually a straightforward generalization of a result of Gangbo [9] which was an important tool in [9] to prove Brenier's Theorem.

Proposition 3 *Let ϕ be $\check{h}$-convex and finite, let $f \in C^0(Y, \mathbb{R})$, define:*

$$\psi_0 := \phi^{\check{h}}$$

and for all $r \in (-1, 1)$

$$\psi_r := (\phi + rf)^{\check{h}}$$

then also define $A := \{x \in \Omega : \psi_0$ is differentiable at $x\}$ and $s := s_{\psi_0}$ as in Proposition 2, 4) then, for all $x \in A$:

$$\lim_{r \to 0} \frac{1}{r}[\psi_r(x) - \psi_0(x)] = -f(s(x)). \tag{9}$$

Since $\mu(\overline{\Omega} \setminus A) \doteq 0$, (9) is satisfied a.e. in $\overline{\Omega}$.

Proof.

Let $x \in A$, first we have:

$$\psi_0(x) = h(x, s(x)) - \phi(s(x)) \tag{10}$$

And, for all $r \in (-1, 1)$:

$$\psi_r(x) = h(x, y_r) - \phi(y_r) - rf(y_r) \text{ for all } y_r \in \partial^h \psi_r(x). \tag{11}$$

Let r_n be some sequence of $(-1, 1) \setminus \{0\}$ which converges to 0 and relabel some sequence $y_{r_n} \in \partial^h \psi_{r_n}(x)$ into y_n.

Step 1.

Let us show first that $y_n \to s(x)$ as $n \to +\infty$.

Up to a subsequence we may assume that y_n converges to $y \in Y$. First note that:

$$\|\psi_{r_n} - \psi_0\|_\infty \leq r_n \|f\|_\infty \to 0 \tag{12}$$

and:

$$\psi_{r_n}(x) = h(x, y_n) - \phi(y_n) - r_n f(y_n). \tag{13}$$

Since ϕ is l.s.c., we get:

$$\underline{\lim}_n \phi(y_n) \geq \phi(y)$$

so that passing to the limit in (13) $\psi_0(x) \leq h(x, y) - \phi(y) = h(x, y) - \psi_0^h(y)$ and then $y \in \partial^h \psi_0(x) = \{s(x)\}$, $s(x)$ is therefore the only cluster point of y_n so that the whole sequence converges to $s(x)$.

Step 2.

First, we have:

$$\frac{1}{r_n}[\psi_{r_n}(x)-\psi_0(x)] = \frac{1}{r_n}[(h(x,y_n)-\phi(y_n))-(h(x,s(x))-\phi(s(x))]-f(y_n). \tag{14}$$

On the one hand:

$$h(x,y_n)-\phi(y_n) \leq h(x,s(x))-\phi(s(x)) \tag{15}$$

on the other hand:

$$h(x,y_n)-\phi(y_n) \geq h(x,s(x))-\phi(s(x))+r_n[f(y_n)-f(s(x))] \tag{16}$$

using (15), (16) and the fact that y_n converges to $s(x)$ and passing to the limit in (14) we obtain:

$$\lim_n \frac{1}{r_n}[\psi_{r_n}(x)-\psi_0(x)] = -f(s(x)) \tag{17}$$

since (17) holds for any sequence $(r_n) \in ((-1,1)\setminus\{0\})^{\mathbb{N}}$ that converges to 0 we finally get:

$$\lim_{r\to 0} \frac{1}{r}[\psi_r(x)-\psi_0(x)] = -f(s(x)). \tag{18}$$

□

3. Proof of the main statement

We are now ready to prove Theorem 1. First note that one obviously has:

$$\sup(\mathcal{MK}) \geq \sup(\mathcal{M}). \tag{19}$$

Let $(\psi,\phi) \in E_h$ and $\gamma \in \Gamma(\mu,\nu)$, one has:

$$L(\psi,\phi) = \int_{\Omega\times Y}(\psi(x)+\phi(y))d\gamma(x,y)$$

$$\geq \int_{\Omega\times Y} h(x,y)d\gamma(x,y) = K(\gamma)$$

so that:

$$\inf(\mathcal{D}) \geq \sup(\mathcal{MK}) \tag{20}$$

Remark. If $(\psi, \phi) \in E_h$ and $s \in \Delta(\mu, \nu)$ (respectively $\gamma \in \Gamma(\mu, \nu)$) are such that $J(s) = L(\psi, \phi)$ (respectively $K(\gamma) = L(\psi, \phi)$) then (ψ, ϕ) is a solution of $(\mathcal{D})$ and s is a solution of $(\mathcal{M})$ (respectively γ is a solution of $(\mathcal{MK})$). —

The first step of the proof is:

Lemma 1 *There exists a solution $(\overline{\psi}, \overline{\phi})$ of $(\mathcal{D})$, moreover if (ψ, ϕ) is a solution of $(\mathcal{D})$ then ψ is h-convex, ϕ is $\check{h}$-convex and those functions are conjugate to each other: $\phi = \psi^h$ ν-a.e. and $\psi = \phi^{\check{h}}$ μ-a.e..*

Proof.

Note first that it is clear from (20) that the value of $(\mathcal{D})$ is finite.

Step 3.

We first prove that if $(\psi, \phi) \in E_h$ is a solution of $(\mathcal{D})$ then:

$$\mu(\{\psi > (\psi^h)^{\check{h}}\}) = \mu(\{\psi > \phi^{\check{h}}\}) = \nu(\{\phi > (\phi^{\check{h}})^h\}) = 0 \tag{21}$$

If $(\psi, \phi) \in E_h$ then obviously $\psi \geq \phi^{\check{h}}$ and $\phi \geq \psi^h$. Let $\widetilde{\psi} := \phi^{\check{h}}$ and $\widetilde{\phi} := \widetilde{\psi}^h = (\phi^{\check{h}})^h$, by Young's inequality $(\widetilde{\psi}, \widetilde{\phi}) \in E_h$ and $\widetilde{\psi} \leq \psi$ and $\widetilde{\phi} \leq \phi$ so that $L(\psi, \phi) \geq L(\widetilde{\psi}, \widetilde{\phi})$. Hence if $(\psi, \phi) \in E_h$ is a solution of $(\mathcal{D})$ then:

$$\nu(\{\phi > (\phi^{\check{h}})^h\}) = 0$$

and

$$\mu(\{\psi > \phi^{\check{h}}\}) = 0$$

this also implies $\mu(\{\psi > (\psi^h)^{\check{h}}\}) = 0$ since $(\psi^h)^{\check{h}} \geq \phi^{\check{h}}$ and (21) is proved.

Step 4.

We now prove existence. Let $(\psi_n, \phi_n) \in E_h^{\mathbb{N}}$ be some minimizing sequence of $(\mathcal{D})$, noting that $L(\psi_n + a, \phi_n - a) = L(\psi_n, \phi_n)$ and using (21), we may assume with no loss of generality that $\psi_n = \phi_n^{\check{h}}$, $\phi_n = \psi_n^h$ and:

$$\inf_Y \phi_n = 0 \tag{22}$$

also note that the infimum in (22) is attained since ϕ_n is l.s.c. say at some point z_n. Since $\phi_n \geq 0$ we get first:

$$\psi_n \leq \max_{\overline{\Omega} \times Y} h$$

and since $\phi_n(z_n) = 0$:

$$\psi_n \geq h(., z_n) \geq \min_{\overline{\Omega} \times Y} h$$

so that $\psi_n(x)$ is bounded uniformly in n and $x \in \Omega$. On the other hand, using the fact that ψ_n is h-convex and Proposition 2, assertion 2), we get that ψ_n is locally Lipschitz uniformly in n. Using Ascoli's Theorem, we may assume, up to a subsequence that ψ_n converges uniformly on compact subsets of Ω to some bounded and locally Lipschitz function $\overline{\psi}$.

Step 5.

Let us prove that $\overline{\psi}$ is itself h-convex. Define, for all $x \in \Omega$:

$$\widetilde{\psi}(x) := \sup_{x' \in \Omega,\ y' \in F(x')} \overline{\psi}(x') + h(x, y') - h(x', y')$$

with:

$$F(x') := \bigcap_{N \geq 1} \overline{\bigcup_{n \geq N} \partial^h \psi_n(x')}$$

note that $F(x') \neq \emptyset$ for $\partial^h \psi_n(x')$ is nonempty and compact for all n.

$\widetilde{\psi}$ is clearly h-convex and $\widetilde{\psi} \geq \overline{\psi}$ let us show the converse inequality: let $x' \in \Omega$ and $y' = \lim_N y_{n_N}$ where $n_N \to +\infty$ and $y_{n_N} \in \partial^h \psi_{n_N}(x')$, passing to the limit in:

$$\psi_{n_N}(x) \geq \psi_{n_N}(x') + h(x, y_{n_N}) - h(x', y_{n_N})$$

we get:

$$\overline{\psi}(x) \geq \overline{\psi}(x') + h(x, y') - h(x', y')$$

taking the supremum in the previous finally proves $\overline{\psi} = \widetilde{\psi}$ so that $\overline{\psi}$ is h-convex.

Step 6.

Let $\overline{\phi} := \overline{\psi}^h$ (so that $(\overline{\psi}, \overline{\phi}) \in E_h$) and let us prove that $(\overline{\psi}, \overline{\phi})$ is a solution of $(\mathcal{D})$. Lebesgue's dominated convergence Theorem yields first:

$$\int_\Omega \psi_n d\mu \to \int_\Omega \overline{\psi} d\mu. \tag{23}$$

Now since, for all $(x, y) \in \Omega \times Y$, $\phi_n(y) \geq h(x, y) - \psi_n(x)$ we get:

$$\underline{\lim}_n \phi_n \geq \overline{\psi}^h = \overline{\phi} \tag{24}$$

using (24) and Fatou's Lemma we get:

$$\underline{\lim}_n \int_Y \phi_n d\nu \geq \int_Y \overline{\phi} d\nu. \tag{25}$$

By (23) and (25) we deduce that $(\overline{\psi}, \overline{\phi})$ is a solution of $(\mathcal{D})$ with $\overline{\psi} = \overline{\phi}^{\check{h}}$ since $\overline{\phi} = \overline{\psi}^h$ and $\overline{\psi}$ is h-convex.

□

The precise duality relations between $(\mathcal{D})$ and $(\mathcal{M})$, $(\mathcal{MK})$ are given by:

Lemma 2 *Let $(\overline{\psi}, \overline{\phi})$ be as in the previous Lemma and $\overline{s}$ be any Borel selection of $\partial^h\overline{\psi}$, the following assertions hold:*

1) $\overline{s} \in \Delta(\mu, \nu)$ and it is a solution of $(\mathcal{M})$,

2) $\overline{\gamma} := (id, \overline{s})\sharp\mu \in \Gamma(\mu, \nu)$ and it is a solution of $(\mathcal{MK})$

3) $(\mathcal{D})$ is dual to $(\mathcal{M})$ and $(\mathcal{MK})$ in the sense:

$$\overline{v} := \inf(\mathcal{D}) = \sup(\mathcal{M}) = \sup(\mathcal{MK}).$$

Proof. Since $(\phi^{\check{h}}, \phi) \in E_h$ for all ϕ, $\overline{\phi}$ minimizes $\phi \mapsto \widetilde{L}(\phi) := L(\phi^{\check{h}}, \phi)$ say for instance in $L^\infty(Y, \mathcal{B}_Y, \nu)$.

In particular for all $f \in C^0(Y, \mathbb{R})$:

$$\lim_{r \to 0+} \frac{1}{r}[\widetilde{L}(\overline{\phi} + rf) - \widetilde{L}(\overline{\phi})] \geq 0 \tag{26}$$

$$\frac{1}{r}[\widetilde{L}(\overline{\phi} + rf) - \widetilde{L}(\overline{\phi})] = \int_Y f d\nu + \int_\Omega \frac{1}{r}[(\overline{\phi} + rf)^{\check{h}} - \overline{\phi}^{\check{h}}]d\mu$$

Proposition 3, yields first:

$$\lim \frac{1}{r}[(\overline{\phi} + rf)^{\check{h}}(x) - \overline{\phi}^{\check{h}}(x)] = -f(\overline{s}(x)), \ \mu\text{-a.e.} \tag{27}$$

on the other hand:

$$|\frac{1}{r}[(\overline{\phi} + rf)^{\check{h}} - \overline{\phi}^{\check{h}}]| \leq \|f\|_\infty \tag{28}$$

(26), (27), (28) and Lebesgue's Dominated Convergence Theorem yield then:

$$\int_Y f d\nu - \int_\Omega f(\overline{s}(x))d\mu(x) \geq 0 \tag{29}$$

and the converse inequality obviously holds changing f into $-f$. To sum up, we have proved:

$$\int_Y f d\nu = \int_\Omega f(\overline{s}(x))d\mu(x) \tag{30}$$

for all $f \in C^0(Y, \mathbb{R})$ so that $\nu = \overline{s}\sharp\mu$. In other words $\overline{s} \in \Delta(\mu, \nu)$ and $\overline{\gamma} := (id, \overline{s})\sharp\mu \in \Gamma(\mu, \nu)$. Now note that since $\overline{\psi}(x) + \overline{\phi}(\overline{s}(x)) = h(x, \overline{s}(x))$ and using $\overline{s}\sharp\mu = \nu$ we have:

$$L(\overline{\psi}, \overline{\phi}) = \inf(\mathcal{D}) = \int_\Omega [\overline{\psi}(x) + \overline{\phi}(\overline{s}(x))]d\mu(x)$$

$$= \int_{\Omega} h(x, \overline{s}(x)) d\mu(x) = J(\overline{s}) = K(\overline{\gamma})$$

Finally, using (20) we get:

$$J(\overline{s}) = \sup(\mathcal{M}) = K(\overline{\gamma}) = \sup(\mathcal{MK}) = \inf(\mathcal{D}) \tag{31}$$

which proves that $\overline{s}$ is a solution of($\mathcal{M}$) and $\overline{\gamma}$ is a solution of ($\mathcal{MK}$), hence the desired result.

□

The last thing to prove is uniqueness:

Lemma 3 *Let $(\overline{\psi}, \overline{\phi})$, $\overline{s}$ and $\overline{\gamma}$ be as in the previous Lemma, the following assertions hold:*

1) if (ψ, ϕ) is a solution of $(\mathcal{D})$ then there exists a constant a such that:

$$\psi - \overline{\psi} = a, \ \mu\text{-a.e.},$$

$$\phi - \overline{\phi} = -a, \ \nu\text{-a.e.},$$

2) if s is a solution of $(\mathcal{M})$ then $s = \overline{s}$ μ-a.e.,

3) $\overline{\gamma} := (id, \overline{s})\sharp\mu$ is the unique solution of $(\mathcal{MK})$.

Proof.

1) Assume that (ψ, ϕ) is a solution of $(\mathcal{D})$ then, using Lemma 1, we may assume that

$$\psi = \phi^{\check{h}} \text{ and } \phi = \psi^h$$

let s be some Borel selection of $\partial^h \psi$. We already know that s is a solution of $(\mathcal{M})$ by Lemma 2. Young's inequality yields:

$$\overline{\psi}(x) + \overline{\phi}(s(x)) \geq h(x, s(x)), \text{ for all } x \in \Omega \tag{32}$$

using $J(s) = L(\overline{\psi}, \overline{\phi})$ and the fact that $s \in \Delta(\mu, \nu)$ we get:

$$L(\overline{\psi}, \overline{\phi}) = \int_{\Omega} [\overline{\psi}(x) + \overline{\phi}(s(x))] d\mu(x) = \int_{\Omega} h(x, s(x)) d\mu(x) \tag{33}$$

(33) and (32) yield:

$$\overline{\psi}(x) + \overline{\phi}(s(x)) = h(x, s(x)), \text{ for } \mu\text{-almost every } x \in \Omega$$

or equivalently $s(x) \in \partial^h \overline{\psi}(x)$ a.e..

Finally, Corollary 1 implies that there exists some constant a

$$\psi - \overline{\psi} = a, \ \mu\text{-a.e. and } \phi - \overline{\phi} = -a \ \nu\text{-a.e.}$$

2) Similarly, if s is a solution of $(\mathcal{M})$ then

$$\overline{\psi}(x) + \overline{\phi}(s(x)) = h(x, s(x)), \text{ for } \mu\text{-almost every } x \in \Omega$$

and then $s = \overline{s}$ μ-a.e.

3) Let γ be a solution of $(\mathcal{MK})$ so that:

$$K(\gamma) = \int_{\Omega \times Y} h d\gamma = L(\overline{\psi}, \overline{\phi}) = \int_{\Omega \times Y} [\overline{\psi}(x) + \overline{\phi}(y)] d\gamma(x, y) \tag{34}$$

since $(\overline{\psi}, \overline{\phi}) \in E_h$ we get:

$$h(x, y) = \overline{\psi}(x) + \overline{\phi}(y), \ \gamma\text{-a.e.} \tag{35}$$

Let $G(\overline{s})$ be the graph of $\overline{s}$ and $G(\partial^h \overline{\psi})$ be that of $\partial^h \overline{\psi}$, (35) implies:

$$\gamma(G(\overline{s})) = \gamma(G(\partial^h \overline{\psi})) = 1. \tag{36}$$

Let A be some Borel subset of Ω and B be some Borel subset of Y, by (36), we get:

$$\gamma(A \times B) = \gamma(A \times B \cap G(\overline{s}))$$

using (36) once again, we get then:

$$\gamma(A \times B) = \gamma((A \cap \overline{s}^{-1}(B)) \times Y)$$

and since $\pi_1 \sharp \gamma = \mu$:

$$\gamma(A \times B) = \mu(A \cap \overline{s}^{-1}(B)) = \overline{\gamma}(A \times B)$$

so that $\gamma = \overline{\gamma}$ which ends the proof.

□

We end this section by a Polar-Factorization type consequence of the main result:

Corollary 2 *Up to μ-a.e. equivalence there exists a unique Borel map s such that:*

1) there exists some h-convex potential ψ such that $s(x) \in \partial^h \psi(x)$ for all $x \in \Omega$,

2) s pushes forward μ through ν.

Proof. $\overline{s}$ defined as previously, satifies the desired result, now, if s satisfies 1) and 2) then

$$J(s) = L(\psi, \psi^h) \geq \inf(\mathcal{D}) = \sup(\mathcal{M})$$

so that, using Lemma 3, $s = \overline{s}$ μ-a.e.

□

4. Economic application and generalized Spence-Mirrlees condition

We end this article by an application of our result to the theory of incentives. More precisely, we are going to prove a re-allocation principle that generalizes a well-known one in dimension 1.

Assume that agents' preferences are given by the quasi-linear utility function:

$$V(x, y, t) = h(x, y) + t,$$

where $x \in \Omega$ is the agent's type or parameter, $y \in Y$ is an action and $t \in \mathbb{R}$ is some monetary compensatory transfer. We make the same assumptions on Ω, Y, h and μ as previously. Note that in this case, the probability measure μ captures the distribution of types among agents.

A key concept in that theory is that of *incentive-compatible contracts*:

Definition 4 *1) A contract is a pair of functions* $(s, t) : \Omega \to Y \times \mathbb{R}$.

2) The potential associated with a contract (s, t) *is the function* $V_{s,t}$ *defined by:*

$$V_{s,t}(x) := h(x, s(x)) + t(x) \textit{ for all } x \in \Omega.$$

3) The contract (s, t) *is incentive-compatible if and only if:*

$$h(x, s(x)) + t(x) \geq h(x, s(x')) + t(x'), \textit{ for all} (x, x') \in \Omega^2. \quad (37)$$

4) A function $s : \Omega \to Y$ *is implementable if and only if there exists* $t : \Omega \to \mathbb{R}$ *such that* (s, t) *is incentive-compatible.*

Remark. The incentive-compatibility condition (37) means that it is optimal for every agent to announce his true parameter.

4.1 The usual Spence-Mirrlees condition and the one dimensional case

In the special one-dimensional case i.e. $\Omega = (a, b)$, $Y = [\alpha, \beta]$ and under the assumption that h is of class C^2 and satisfies the Spence-Mirrlees condition:

$$\frac{\partial^2 h}{\partial x \partial y} > 0 \tag{38}$$

then we have the standard characterization result (see [17], [24], [21]):

Proposition 4 *s is implementable if and only if s is nondecreasing.*

Remark. Note that Spence-Mirrlees condition (38) implies our assumption (3) on h.

In this one-dimensional case and under assumption (38), we also have the other characterization:

Lemma 4 *The following assertions are equivalent:*

1) s is nondecreasing

2) there exists ψ h-convex such that $s(x) \in \partial^h \psi(x)$ for all $x \in \Omega$.

Proof. First assume that s is nondecreasing and define:

$$\psi(x) := \int_a^x \frac{\partial h}{\partial x}(t, s(t))dt$$

we are going to prove that ψ is h-convex and $s(x) \in \partial^h \psi(x)$ for all x. Define for all x:

$$\widetilde{\psi}(x) := \sup_{x' \in \Omega} \psi(x') + h(x, s(x')) - h(x', s(x'))$$

$\widetilde{\psi}$ is h-convex and $\widetilde{\psi} \geq \psi$, let us show that $\psi \geq \widetilde{\psi}$. Let $x' \in \Omega$, we have:

$$\psi(x) - \psi(x') - h(x, s(x')) + h(x', s(x'))$$

$$= \int_x^{x'} [\frac{\partial h}{\partial x}(t, s(x')) - \frac{\partial h}{\partial x}(t, s(t))]dt$$

and the latter is nonnegative using the fact that s is nondecreasing and (38). This yields $\psi = \widetilde{\psi}$ so that ψ is h-convex and the previous computation also yields $s(x) \in \partial^h \psi(x)$.

Conversely assume that for all x, $s(x) \in \partial^h \psi(x)$ for some h-convex ψ. We have:

$$\psi(x') - \psi(x) \geq h(x', s(x)) - h(x, s(x))$$

and

$$\psi(x) - \psi(x') \geq h(x, s(x')) - h(x', s(x'))$$

so that:

$$h(x', s(x)) - h(x, s(x)) + h(x, s(x')) - h(x', s(x')) \leq 0$$

or equivalently:

$$\int_x^{x'} [\frac{\partial h}{\partial x}(t, s(x)) - \frac{\partial h}{\partial x}(t, s(x'))]dt \leq 0$$

note finally that, with (38), the previous expression has the sign of $(x' - x)(s(x) - s(x'))$ so that s is nondecreasing.

□

The previous characterizations can be viewed equivalently as a re-allocation principle via monotone rearrangements:

Proposition 5 *Let s_0 be some Borel map : $\Omega \to Y$ and let $\widetilde{s}$ be the non decreasing rearrangement of s_0 with respect to μ then $\widetilde{s}$ is the only Borel map which satisfies:*

1) $\widetilde{s}$ is implementable,

2) $\widetilde{s}$ and s_0 are equimeasurable i.e. : $\widetilde{s}\sharp\mu = s_0\sharp\mu$.

For properties of monotone rearrangements see [18]. Recall that $\widetilde{s}$ is defined as follows; first define:

$$F_{s_0}(y) := \mu(\{s_0 < y\}), \text{ for all } y \in Y$$

then, for all $x \in \Omega$:

$$\widetilde{s}(x) := \inf\{y \in Y \text{ s.t. } F_{s_0}(y) > x\}.$$

Remark. $\widetilde{s}$ is the solution of:

$$\sup_{s \in \Delta(\mu, s_0\sharp\mu)} \int_\Omega h(x, s(x))d\mu(x)$$

In other words, $\widetilde{s}$ maximizes the average surplus in the class of maps that have the same cumulative function as s_0.

The previous remark can be viewed as an easy consequence of Hardy-Littlewood inequality (see [18]).

Let us finally note that if h and g both satisfy (38) then the associated Monge's problems have the same solution. Of course, this result is very specific of the one-dimensional problem.

Proposition 6 *Let h and g be two functions from $\Omega \times Y$ to $\mathbb{R}$ of class C^2 that both satisfy (38); let ν be some Radon probability measure on Y. Then both problems:*

$$\sup_{s \in \Delta(\mu,\nu)} J_h(s) := \int_\Omega h(x, s(x)) d\mu(x)$$

and

$$\sup_{s \in \Delta(\mu,\nu)} J_g(s) := \int_\Omega g(x, s(x)) d\mu(x)$$

have the same solution.

Proof. Let s be the maximizer of J_h over $\Delta(\mu, \nu)$, from Lemma 4 and Theorem 1, s is nondecreasing. There exists then ψ g-convex such that $s(x) \in \partial^g \psi(x)$ for all x and since $s \sharp \mu = \nu$, s maximizes J_g over $\Delta(\mu, \nu)$. □

4.2 Re-allocation principle in the general case

Our aim now is to consider the general problem where Ω is a bounded connected open subset of $\mathbb{R}^n$, and μ is some probability measure in Ω which is absolutely continuous with a positive Radon-Nikodym derivative with respect to the n-dimensional Lebesgue measure, and such that $\mu(\partial\Omega) = 0$, Y is a compact Polish space and h satisfies (1), (2) and (3).

We shall prove a similar re-allocation principle as in the one dimensional case so that (3) is a natural generalization of the Spence-Mirrlees condition. A first attempt was made by Mc Afee and Mc Millan in [15] to characterize incentive-compatibility in a multi-dimensional setting, the condition of these authors is much stronger than (3) and their characterization requires s to be differentiable which is of course not required in what follows since Y need not have a linear structure.

We start by a characterization result that can be found in [5]

Proposition 7 *Let $s : \Omega \to Y$ and $t : \Omega \to \mathbb{R}$ then we have:*

1) (s, t) is incentive-compatible if and only if

$$V_{s,t} \text{ is } h\text{-convex, and}$$

$$s(x) \in \partial^h V_{s,t}(x) \text{ for all } x \in \Omega.$$

2) s is implementable if and only if there exists some h-convex function ψ such that:

$$s(x) \in \partial^h \psi(x) \text{ for all } x \in \Omega.$$

Then the re-allocation principle can be stated as:

Theorem 2 *Let s_0 be an arbitrary Borel function $\Omega \to Y$, there exists a unique (up to a.e. equivalence) Borel map $\overline{s}$ such that:*

1) $\overline{s}$ is implementable,

2) s_0 and $\overline{s}$ are equimeasurable i.e. $\overline{s}\sharp\mu = s_0\sharp\mu$.

Moreover, $\overline{s}$ is the solution of the Monge's Problem:

$$\sup_{s \in \Delta(\mu, s_0\sharp\mu)} \int_\Omega h(x, s(x)) d\mu(x). \tag{39}$$

Proof. Proof follows directly from Theorem 1 and Proposition 7. □

Remark. The economic interpretation of this result is the following : any allocation plan can be rearranged into some implementable one in a unique way ; $\overline{s}$ is therefore in some sense a monotone rearrangement of s_0 and it is obtained by maximizing the average surplus in the set of measure-preserving maps $\Delta(\mu, s_0\sharp\mu)$.

Moreover, at least from a theoretical point of view, one can use our main result to find a tarif $\overline{t}$ such that the pair $(\overline{s}, \overline{t})$ is incentive compatible. Let $(\overline{\psi}, \overline{\phi})$ be a solution of the dual problem of (39), and define for all $x \in \Omega$:

$$\overline{t}(x) := -\overline{\phi}(\overline{s}(x)) = -h(x, \overline{s}(x)) + \overline{\psi}(x),$$

then it can be checked easily that the pair $(\overline{s}, \overline{t})$ is an incentive-compatible contract, let us indeed consider a pair of types (x, x'), we have:

$$h(x, \overline{s}(x)) + \overline{t}(x) = \overline{\psi}(x) = \overline{\phi}^h(x) \geq h(x, \overline{s}(x')) - \overline{\phi}(\overline{s}(x'))$$

$$= h(x, \overline{s}(x')) + \overline{t}(x').$$

References

[1] Aleksandrov, A.D.: Existence and uniqueness of a convex surface with a given integral curvature. C.R. (Doklady) Acad. Sci. URSS **35**, 947-972 (1942)

[2] Benamou, J.D., Brenier, Y.: A numerical method for the optimal time-continuous mass transport problem and related problems. Contemporary Mathematics **226**, 1-11 (1999)

[3] Brenier, Y.: Polar factorization and monotone rearrangements of vector valued functions. Communications in Pure and Applied Mathematics **44**, 375-417 (1991)

[4] Bouchitté, G., Buttazzo, G., Seppecher, P.: Shape optimization solutions via Monge-Kantorovich equation. C.R. Acad. Sci. Paris, 1185-1191 (1998)

[5] Carlier, G.: A general existence result for the principal-agent problem with adverse selection. Journal of Mathematical Economics **35**, 129-150 (2001)

[6] Castaing, C., Valadier, M.: Convex analysis and measurable multifunctions. Lecture Notes in Mathematics **580**, Springer-Verlag 1977

[7] Choné, P.: Étude de quelques problèmes variationnels intervenant en géométrie riemannienne et en économie mathématique.PhD Thesis, Univ. Toulouse I (1999)

[8] Ekeland, I., Temam, R.: Convex Analysis and Variational Problems. North-Holland 1972

[9] Gangbo, W.: An elementary proof of the polar factorization of vector-valued functions. Arch. Rational Mech. Anal. **128**, 381-399 (1994)

[10] Gangbo, W., Mc Cann, R.J.: The geometry of optimal transportation. Acta Math. **177**, 113-161 (1996)

[11] Kantorovich, L.V., Akilov, G.P.: Functional Analysis. Pergamon Press, Oxford 1982

[12] Kantorovich, L.V., Rubinstein, G.S.: On a functional space and certain extremal problems. Dokl. Akad. Nauk. SSSR **115**, 1058-1061 (1957)

[13] Levin, V.: Abstract cyclical monotonicity and Monge solutions for the general Monge-Kantorovich problem. Set-Valued Analysis **7**, 7-32 (1999)

[14] Levin, V.: Reduced cost functions and their applications. Journal of Mathematical Economics **28**, 155-186 (1997)

[15] Mc Afee, R.P., Mc Millan, J.: Multidimensional incentive compatibility and mechanism design. Journal of Economic Theory **46**, 1451-1459 (1988)

[16] Mc Cann, R.J.: Existence and uniqueness of monotone measure-preserving maps. Duke Math. Journal **80**, 309-323 (1995)

[17] Mirrlees, J.: Optimal tax theory : a synthesis. Journal of Public Economics **7**, 327-358 (1976)

[18] Mossino, J.: Inégalités isopérimétriques et applications en physique. Hermann, Paris (1984)

[19] Rachev, S.T.: The Monge-Kantorovich mass transference problem and its stochastic applications. Theory Proba. Appl. **29**, 647-676 (1984)

[20] Rachev, S.T., Rüschendorf, L.: Mass Transportation Problems. Vol. I: Theory; Vol. II : Applications. Springer-Verlag 1998

[21] Rochet, J.-C.: A necessary and sufficient condition for rationalizability in a quasilinear context. Journal of Mathematical Economics **16**, 191-200 (1987)

[22] Rochet, J.-C., Choné, P.: Ironing, sweeping and multidimensional screening. Econometrica **66**, 783-826 (1998)

[23] Rockafellar, R.T.: Convex Analysis. Princeton University Press 1970

[24] Spence, M.: Competitive and optimal responses to signals. Journal of Economic Theory **7**, 296-332 (1974)
[25] Wagner, D.H.: Survey of measurable selection theorems : an update. In: Measure Theory. Lecture Notes in Math. **794**, pp.176-219 Springer-Verlag 1980

Adv. Math. Econ. 5, 23–54 (2003)

Advances in
MATHEMATICAL ECONOMICS

Functional evolution equations governed by m-accretive operators

Charles Castaing[1] **and Ahmed Gamal Ibrahim**[2]

[1] Département de Mathématiques, Case 051, Université Montpellier II, F-34095 Montpellier cedex 5, France
[2] Department of Mathematics, Faculty of Sciences, Cairo University, Cairo, Egypt

Received: September 1, 2002
Revised: December 1, 2002

JEL classification: C62, C72

Mathematics Subject Classification (2000): 28A33, 28A50

Abstract. Let $r > 0$ be a finite delay and $\mathcal{C}_0 = \mathcal{C}_E([-r, 0])$ be the Banach space of continuous vector-valued functions defined on $[-r, 0]$ taking values in a separable reflexive Banach space E such that its strong dual E' is uniformly convex. We discuss the existence of strong solutions for an evolution equation of the form $\dot{u}(t) \in -A(t)u(t) + F(t, \tau(t)u)$ a.e in $[0, 1]$, where $u : [-r, 1] \to E$ is a continuous mapping from $[-r, 1]$ into E such that its restriction to $[-r, 0]$ is equal to $\varphi \in \mathcal{C}_0$, and its restriction to $[0, 1]$ is absolutely continuous, (i.e $u(t) = u(0) + \int_0^t \dot{u}(s)\, ds, \forall t \in [0, 1]$ with $\dot{u} \in L^1_E([0, 1])$), and satisfies the preceding inclusion, $A(t)$ is an m-accretive multivalued operator on E, $F : [0, 1] \times \mathcal{C}_0 \to E$ is a convex weakly compact valued, separately scalarly measurable and separately scalarly upper semicontinuous multifunction and $(\tau(t)u)(s) = u(t + s), \forall s \in [-r, 0]$. Some applications to the sweeping process (or Moreau's process) and Optimal Control involving Young measures are also presented.

Key words: accretive operator, ball-compact, functional differential inclusion, maximal monotone operator, multifunction, original control, relaxed control, relaxation, Young measure

Introduction

Let $r > 0$ be a finite delay, E be a reflexive separable Banach space such that its strong dual E' is uniformly convex, $A(t), (t \in [0, 1])$ be an m-accretive multivalued operator on $E, \mathcal{C}_0 = \mathcal{C}_E([-r, 0])$ be the Banach space of all continuous E-valued functions defined on $[-r, 0]$ equipped with the norm of uniform convergence and $F : [0,1] \times \mathcal{C}_E([-r,0]) \to E$ be a separately scalarly measurable and separately scalarly upper semicontinuous convex weakly compact valued multifunction. For any $t \in [0,1]$, let $\tau(t) : \mathcal{C}_E([-r,t]) \to \mathcal{C}_0$ defined by $(\tau(t)u)(s) = u(t+s), \forall s \in [-r,0]$ and $\forall u \in \mathcal{C}_E([-r,t])$. Let φ be a given element of $\mathcal{C}_0$. This paper concerns the existence of strong solutions for an evolution equation governed by m-accretive operators of the type

$$(P_\tau) \begin{cases} \dot{u}(t) \in -A(t)u(t) + F(t, \tau(t)u), \ a.e\ t \in [0,1], \\ u(s) = \varphi(s), \ \forall s \in [-r,0]; \ u(t) \in \overline{D(A(t))}, \ \forall t \in [0,1]. \end{cases}$$

By a strong solution of (P_τ) we mean a function $u : [-r,1] \to E$ such that its restriction to $[-r,0]$ is equal to φ and its restriction to $[0,1]$ is absolutely continuous, i.e $u(t) = u(0) + \int_0^t \dot{u}(s)\, ds, \forall t \in [0,1]$ with $\dot{u} \in L^1_E([0,1])$ and satisfies (P_τ).

In section 2 we state the existence of strong solutions for the problem (P_τ) and its consequences via a new discretization technique. As an illustration of this technique, we present also a viable strong solution to an evolution equation governed by an m-accretive operator $A(t)$ of the form $\dot{u}(t) \in -A(t)u(t)$ $a.e\ t \in [0,1]; u(0) = a \in C(0) \cap \overline{D}$, $u(t) \in C(t) \cap \overline{D}$, $\forall t \in [0,1]$, here $\overline{D} = \overline{D(A(t))}$, for all $t \in [0,1]$, and $C : [0,1] \to E$ is a closed *ball-compact* valued (that is the intersection of $C(t)$ with any closed ball of E is compact), upper semicontinuous multifunction, satisfying a viablity type condition, namely $[I_E + \lambda A(t)]^{-1} C(s) \subset C(t)$, for all $\lambda > 0$ and for all $0 \le s \le t \le 1$. Applications to sweeping process [7, 8, 20, 21, 26, 27] and Optimal Control problems for the equations under consideration with the help of Young measures (see e.g [28]) are given in section 3. The present work is essentially a continuation of the ones given in [4, 5, 7, 8, 13, 15, 25, 26]. For more on evolution problems with delay in Banach spaces and related results, see [18, 23, 24, 31] and the references therein.

1. Notations and preliminaries

We will use the following definitions and notations and summarize some basic results.

- E is a separable Banach space, I_E is the identity mapping on E and E' is the topological dual of E.
- $\bar{B}_E$ is the closed unit ball of E.
- $cc(E)$ (resp. $cwk(E)$) is the collection of nonempty closed convex (resp. weakly compact convex) subsets of E.
- If A is a subset of E, $\delta^*(., A)$ is the support function of A.

- $\mathcal{L}([0,1])$ is the σ-algebra of Lebesgue measurable subsets of $[0,1]$.
- If X is a topological space, $\mathcal{B}(X)$ is the Borel tribe of X.
- If A and B are two subsets of E, the excess of A over B is

$$e(A,B) = \sup\{d(a,B) :\ a \in A\}$$

and its Hausdorff distance is

$$h(A,B) = \sup\{e(A,B), e(B,A)\}.$$

The excess $e(A,\{0\})$ is denoted by $|A|$ where $|A| = \sup\{||a|| : a \in A\}$.

- $L^1_E([0,1],dt)$ (shortly $L^1_E([0,1])$) is the Banach space of Lebesgue-Bochner integrable functions $f : [0,1] \to E$.
- A mapping $u : [0,1] \to E$ is *absolutely continuous* if there is a function $\dot{u} \in L^1_E([0,1])$ such that $u(t) = u(0) + \int_0^t \dot{u}(s)\,ds,\ \forall t \in [0,1]$.
- If X is a topological space, $C_E(X)$ is the space of continuous mappings $u : X \to E$ equipped with the norm of uniform convergence: $||u||_{C_E(X)} = \sup_{x \in X} ||u(x)||$.
- A multifunction $F : [0,1] \to \mathcal{B}(E)$ is measurable if its graph belongs to $\mathcal{L}([0,1]) \otimes \mathcal{B}(E)$. A convex weakly compact valued multifunction $F : X \to cwk(E)$ defined a topological space X is scalarly upper semicontinuous if for every $x' \in E'$, the scalar function $\delta^*(x', F(.))$ is upper semicontinuous on X.
- A multivalued operator $A(t) : E \to 2^E$, $(t \in [0,1])$ is *m-accretive*, if, for each $t \in [0,1]$ and each $\lambda > 0$, $R(I_E + \lambda A(t)) = E$, and for each $x_1 \in D(A(t)), x_2 \in D(A(t)), y_1 \in A(t)x_1, y_2 \in A(t)x_2$, we have

$$(j) \qquad ||x_1 - x_2|| \leq ||(x_1 - x_2) + \lambda(y_1 - y_2)||$$

where $D(A(t)) := \{x \in E : A(t)x \neq \emptyset\}$.

If $A(t)$ is m-accretive, then

$$(jj) \qquad \frac{1}{\lambda}||J_\lambda A(t)x - x|| = ||A_\lambda(t)x|| \leq |A(t)x|_0 := \inf_{y \in A(t)x} ||y||, \quad \forall x \in D(A(t)),$$

where $J_\lambda A(t)x = (I_E + \lambda A(t))^{-1}x$. We refer to [1, 3, 31] for the theory of accretive operators and equations of evolution in Banach spaces. We refer to [9] for Measurable Multifunctions and Convex Analysis.

Throughout E is a separable reflexive Banach space such that its strong dual is uniformly convex.

2. Existence results

We will consider the following assumptions:

(H_1) There exist a continuous function $\rho : [0,1] \to E$ and a nondecreasing function $L : [0,\infty[\to [0,\infty[$ such that

$$||J_\lambda A(t)x - J_\lambda A(s)x|| \leq \lambda|\rho(t) - \rho(s)|L(||x||)$$

for all $\lambda \in]0,1]$, for all $(t,s) \in [0,1] \times [0,1]$, and for all $x \in E$.

(H_2) For every $s > 0$, there exists $\delta(s) > 0$ such that

$$||J_\lambda A(0)x - x|| \leq \lambda\delta(s)$$

for all $\lambda \in]0,1]$ and for all $x \in \overline{D(A(0))}$ with $||x|| \leq s$.

Remarks. 2.1. Assumption (H_1) is similar to the one employed by [3, 12] in the study of quasi-autonomuous evolution equations. By ([12], Lemma 3.1), (H_1) implies that the sets $D(A(t))$ are constant, i.e $D(A(t)) := D$ for all $t \in [0,1]$.

2.2. In view of (jj) (H_2) is satisfied if, $0 \in D(A(0)) = \overline{D(A(0))}$ and $A(0)$ satisfies the following boundness type condition, namely, for any closed ball $\overline{B}(0,\eta)$ of center 0 with radius η, the set $\{|A(0)x|_0 : x \in D(A(0)) \cap \overline{B}(0,\eta)\}$ is bounded in $\mathbb{R}$. In particular, (H_2) is satisfied if $A(0) : \underline{D(A(0))} \to E$ is $cwk(E)$-valued and scalarly upper semicontinuous, and $D(A(0)) \cap \overline{B}(0,\eta)$ is compact for any closed ball $\overline{B}(0,\eta)$ of center 0 with radius η, because the sets $\{A(0)x : x \in D(A(0)) \cap \overline{B}(0,\eta)\}$ are therefore weakly compact.

(H_3) (a) For every $L^2_E([0,1])$-mapping $u : [0,1] \to E$ satisfying $u(t) \in D(A(t))$ for all $t \in [0,1]$, the multifunction $t \to A(t)u(t)$ is measurable, (b) for every $x \in E$ and for every $\lambda > 0$, $t \mapsto (I_E + \lambda A(t))^{-1}x$ is measurable, (c) there exists $\overline{g} \in L^2_E([0,1])$ such that $t \mapsto (I_E + \lambda A(t))^{-1}\overline{g}(t)$ belongs to $L^2_E([0,1])$ for all $\lambda > 0$.

(H_4) $F : [0,1] \times C_E([-r,0]) \to cwk(E)$ is separately scalarly measurable on $[0,1]$ and separately scalarly upper semicontinuous on $C_E([-r,0])$, and there exists a convex weakly compact set K such that $F(t,u) \subset K$ for all $(t,u) \in [0,1] \times C_E([-r,0])$.

The following provides a closure type result in the convergence of approximated solutions.

Lemma 2.3. *Suppose that $A(t) : E \to 2^E (t \in [0,1])$, is an m-accretive operator satisfying $(H_3), (u_n)$ and (v_n) are sequences in $L^2_E([0,1])$ with*

$u_n(t) \in D(A(t))$ for every n and for every $t \in [0,1]$ and (r_n) is a uniformly bounded sequence of positive measurable functions defined on $[0,1]$ such that $r_n(t) \to 0$ pointwisely on $[0,1]$. Assume that the following conditions are satisfied:

(i) (u_n) converges strongly to $u \in L^2_E([0,1])$ and (v_n) converges to $v \in L^2_E([0,1])$ with respect to the topology $\sigma(L^2_E, L^2_{E'})$,

(ii) $v_n(t) \in A(t)u_n(t) + r_n(t)\overline{B}_E$ for all n and all $t \in [0,1]$.

Then we have $v(t) \in A(t)u(t)$ a.e $t \in [0,1]$.

Proof. Let $I_{L^2_E([0,1])}$ be the identity operator in $L^2_E([0,1])$. Let $\mathcal{A}$ be the operator in $L^2_E([0,1])$ defined by

$$v \in \mathcal{A}u \iff v(t) \in A(t)u(t) \ a.e \ t \in [0,1].$$

We claim that $\mathcal{A}$ is m-accretive in $L^2_E([0,1])$. Let $\lambda > 0$ and let $g \in L^2_E([0,1])$. By $(H_3)(c)$ there exists $\overline{g} \in L^2_E([0,1])$ such that $\overline{h} : t \mapsto (I_E + \lambda A(t))^{-1}\overline{g}(t)$ belongs to $L^2_E([0,1])$. Since $(I_E + \lambda A(t))^{-1}$ is nonexpansive [31], we deduce that the function $h : t \mapsto (I_E + \lambda A(t))^{-1}g(t)$ is measurable and belongs to $L^2_E([0,1])$ thanks to $(H_3)(b)$-(c). Furthermore, we have $g \in h + \lambda\mathcal{A}h \iff h \in (I_{L^2_E} + \lambda\mathcal{A})^{-1}g \implies R(I_{L^2_E} + \lambda\mathcal{A}) = L^2_E$. Let $\mathcal{U}$ be the closed unit ball of $L^\infty_E([0,1])$. In view of (ii), $(H_3)(a)$ and measurable selection theorem, we claim that $v_n \in \mathcal{A}u_n + \mathcal{R}_n$ for all n, where

$$\mathcal{R}_n := \{z \in L^\infty_E([0,1]) : z = r_n w, \ w \in \mathcal{U}\}.$$

Firstly, it is easy to see that $\mathcal{R}_n$ is equal to the set of all measurable selections of the measurable multifunction $r_n(.)\overline{B}_E$. Secondly, by (ii) and $(H_3)(a)$, the nonempty-valued multifunction $\Psi_n : [0,1] \to E \times E$ defined by

$$\Psi_n(t) := \{(x,y) \in (A(t)u_n(t), r_n(t)\overline{B}_E) : x + y = v_n(t)\}, \ \forall t \in [0,1],$$

is measurable. By measurable selection theorem, there is a measurable selection x_n of the multifunction $A(.)u_n(.)$ and a measurable selection y_n of the measurable multifunction $r_n(.)\overline{B}_E$ such that $v_n(t) = x_n(t) + y_n(t)$ for all $t \in [0,1]$. Moreover there is $w_n \in \mathcal{U}$ such that $y_n = r_n w_n$. So, we have that $v_n = x_n + r_n w_n \in \mathcal{A}u_n + \mathcal{R}_n$ for all n. As $A(t)$ is accretive for each $t \in [0,1]$, it is easy to check that $\mathcal{A}$ is accretive in $L^2_E([0,1])$. Since E' is uniformly convex, the dual $L^2_{E'}([0,1])$ of $L^2_E([0,1])$ is uniformly convex, too, see e.g ([32], Theorem 4.2 and Remark 4.7). Consequently, by ([31], Theor.1.5.2) the graph of $\mathcal{A}$ is strongly-weakly sequentially closed. By (i) u_n strongly converges to $u \in L^2_E([0,1])$, $v_n - r_n w_n \to v$, weakly in $L^2_E([0,1])$, (because $r_n w_n \to 0$ strongly in $L^2_E([0,1])$ and $v_n - r_n w_n \in \mathcal{A}u_n$ by what has been proved, so we conclude that $v \in \mathcal{A}u \iff v(t) \in A(t)u(t) \ a.e \ t \in [0,1]$. □

Remarks. (1) If $r_n(.) = 0$, $\forall n$, $H_3(a)$ can be dropped. In particular, Lemma 2.3 shows that the m-accretiveness lifts from E to $L^2_E([0,1])$.

(2) Here is an example of a maximal monotone operator in a separable Hilbert space H satisfying (H_3). Let $f : [0,1] \times H \to \mathbb{R}$ be a Carathéodory function such that $f(t,.)$ is convex and such that $|f(t,x) - f(t,y)| \leq \beta(t)||x - y||$, for all $(t,x,y) \in [0,1] \times H \times H$ with $\beta \in L^2_{\mathbb{R}}([0,1])$. Let ∂f_t denotes the subdifferential of f_t. Then the following hold. For every $v \in H$, and for every $x \in H$, the subdifferential $\partial f_t(x)$ of f_t at the point x is a convex weakly compact subset of H and its the support function $\delta^*(v, \partial f_t(x))$ is equal to

$$f'_t(x; v) = \inf_{\delta>0} \frac{f_t(x + \delta v) - f_t(x)}{\delta}.$$

Taking the expression of $f'_t(x;v)$ into account, it is easy to check that, if $x : [0,1] \to H$ and $y : [0,1] \to H$ are two measurable mappings, then the function $t \mapsto f'_t(x(t); y(t))$ is measurable. In particular this shows that the multifunction $t \to A(t)x(t) := \partial f_t(x(t))$ is scalarly measurable, hence $t \to A(t)x(t)$ is measurable. So $A(t)$ satisfies $(H_3)(a)$. Let $g : [0,1] \to H$ be a measurable mapping. We claim that the single-valued function $t \to (I_H + \lambda A(t))^{-1}g(t)$ is measurable. First suppose that g is continuous and f is globally continuous. It is clear that

$$\begin{aligned}(I_H + \lambda A(t))^{-1}g(t) &= \{y \in H : g(t) \in y + \lambda A(t)y\}\\ &= \{y \in H : d(g(t) - y, \lambda A(t)y) = 0\}.\end{aligned}$$

Notice that, for every $z \in H$

$$d(z, \lambda A(t)y) = \sup_{||x'||\leq 1} [\langle x', z\rangle - \delta^*(x', \lambda A(t)y)].$$

Since the functions

$$(t, y, z) \mapsto \langle x', z\rangle - \delta^*(x', \lambda A(t)y)$$

are lower semicontinuous, because

$$(t, y) \mapsto \delta^*(x', A(t)y) = \inf_{\delta>0} \frac{f_t(y + \delta x') - f_t(y)}{\delta}$$

is upper semicontinuous, so is the function

$$(t, y, z) \mapsto d(z, \lambda A(t)y).$$

Hence

$$\{(t, y) \in [0,1] \times H : d(g(t) - y, \lambda A(t)y) = 0\}$$

is a Borel set in $[0,1] \times H$. Thus $t \mapsto J_\lambda A(t)g(t) = (I_H + \lambda A(t))^{-1}g(t)$ is a Borel function on $[0,1]$ when g is continuous. Consequently $t \to (I_H + \lambda A(t))^{-1}g(t)$ is measurable when g measurable, by applying Lusin's theorem to g and Scorza-Dragoni's theorem to f. So $A(t)$ satisfies $(H_3)(b)$. As the directional derivative $f'_t(x;v)$ of f_t at the point x in the direction v is equal to the support function $\delta^*(v, \partial f_t(x))$ of ∂f_t at the point v, using inequality $|f(t,x) - f(t,y)| \leq \beta(t)||x-y||$ yields

$$\delta^*(v, \partial f_t(x)) = f'_t(x;v) = \inf_{\delta>0} \frac{f_t(x+\delta v) - f_t(x)}{\delta} \leq \beta(t)||v||,$$

so that $\partial f_t(x) = A(t)x \subset \beta(t)\overline{B}_H$ for all $(t,x) \in [0,1] \times H$. Let $A_\lambda(t)$ be the Yosida's approximation of $A(t)$ given by

$$A_\lambda(t) = \frac{1}{\lambda}[I_H - J_\lambda A(t)].$$

By what has been demonstrated, the functions $t \mapsto A_\lambda(t)g(t)$ are measurable when $g(.)$ is measurable. Recall that

$$\begin{aligned} ||A_\lambda(t)x|| &\leq \inf\{||y|| : y \in \partial f_t(x)\} \\ &\leq \sup\{||y|| : y \in \beta(t)\overline{B}_H\} \\ &= \beta(t) \end{aligned}$$

for all $(t,x) \in [0,1] \times H$. Hence the operator subdifferential $A(t) = \partial f_t$ satisfies $(H_3)(a),(b),(c)$. In particular, let us consider a closed convex 1-Lipschitzean moving set $C(t)$ in H

$$|d(x, C(t)) - d(y, C(s))| \leq ||x-y|| + |t-s|$$

for all $x, y \in H$ and for all $t, s \in [0,1]$. Then the subdifferential $\partial d_{C(t)}$ of the distance function $d_{C(t)} : x \mapsto d(x, C(t))$ is a maximal monotone operator depending on t and satisfying (H_3). The preceding example will be useful in the next section. Another example ensuring the measurability of $(I_H + \lambda A(t))^{-1}g(t)$ is in [22] where $A(t) = \partial f_t$ is the subdifferential of a lower semicontinuous convex integrand $f_t (t \in [0,1])$ defined on a Hilbert space.

Theorem 2.4. *Suppose that $A(t) : E \to cc(E) \cup \{\emptyset\}; t \in [0,1]$, is an m-accretive operator, $(H_1),(H_2),(H_3),(H_4)$ are satisfied, and $\overline{D}$ is ball-compact, i.e the intersection of $\overline{D}$ with any closed ball of E is compact. Then for any $\varphi \in C_E([-r,0])$ with $\varphi(0) \in \overline{D}$, there exist at least a strong solution $u \in C_E([-r,1])$ to the problem*

$$(P_\tau) \begin{cases} \dot{u}(t) \in -A(t)u(t) + F(t, \tau(t)u) \; a.e \; t \in [0,1], \\ u(s) = \varphi(s), \; \forall s \in [-r,0]; \; u(t) \in \overline{D}, \; \forall t \in [0,1]. \end{cases}$$

Proof. Step 1. We will prove the theorem in the particular case when (H_4) is replaced by a stronger one: $F : [0,1] \times \mathcal{C}_E([-r,0]) \to cwk(E)$ is scalarly upper semicontinuous on $[0,1] \times \mathcal{C}_E([-r,0])$ and there exists a convex weakly compact set $K \subset E$ such that $F(t,u) \subset K$ for all $(t,u) \in [0,1] \times \mathcal{C}_E([-r,0])$.

We define a sequence of continuous function (u_n) in $\mathcal{C}_E([-r,1])$ such that a subsequence of its converges uniformly to a strong solution of (P_τ). Let $n \geq 1$. We put $u_n = \varphi$ on $[-r,0]$. As F is scalarly upper semicontinuous on $[0,1] \times \mathcal{C}_E([-r,0])$, F is, a fortiori, scalarly $\mathcal{L}([0,1]) \otimes \mathcal{B}(\mathcal{C}_E([-r,0]))$-measurable. Let $\sigma : [0,1] \times \mathcal{C}_E([-r,0]) \to E$ be a scalarly $\mathcal{L}([0,1]) \otimes \mathcal{B}(\mathcal{C}_E([-r,0]))$-measurable selection of F. Let $n \geq 1$ be a fixed integer. In order to construct u_n on $[0,1]$, we consider a partition of $[0,1]$ by the points $t_k^n = ke_n$, $e_n = \frac{1}{n}$, $k = 0,1,2,...,n$. We define u_n by linear interpolation, where $u_n(t_k^n) = x_k^n \in \overline{D(A(t_k^n))}$, $k = 0,1,2,...,n$ are obtained by induction starting with $u_n(0) = x_0^n = \varphi(0) \in \overline{D}$. Let us set

$$x_1^n := J_{e_n}A(t_1^n)(x_0^n + e_n\sigma(t_0^n, \varphi)).$$

By construction we have $x_1^n \in D(A(t_1^n)) \subset \overline{D}$. For $t \in]t_0^n, t_1^n]$, let us set

$$u_n(t) = \frac{t_1^n - t}{e_n} x_0^n + \frac{t - t_0^n}{e_n} x_1^n.$$

Then for $t \in]t_0^n, t_1^n[$, we have

$$\dot{u}_n(t) = \frac{x_1^n - x_0^n}{e_n} \in -A(t_1^n)x_1^n + \sigma(t_0^n, \varphi).$$

By induction on $0 \leq k \leq n$ we set

$$x_{k+1}^n := J_{e_n}A(t_{k+1}^n)(x_k^n + e_n\sigma(t_k^n, \tau(t_k^n)u_n)). \tag{2.4.1}$$

Then $x_{k+1}^n \in D(A(t_{k+1}^n)) \subset \overline{D}$ for $k = 0,1,2,...,n$. For $]t_k^n, t_{k+1}^n]$, $0 \leq k \leq n-1$, we set

$$u_n(t) = \frac{t_{k+1}^n - t}{e_n} x_k^n + \frac{t - t_k^n}{e_n} x_{k+1}^n,$$

then, for $]t_k^n, t_{k+1}^n[$, we have

$$\dot{u}_n(t) = \frac{x_{k+1}^n - x_k^n}{e_n} \in -A(t_{k+1}^n)x_{k+1}^n + \sigma(t_k^n, \tau(t_k^n)u_n). \tag{2.4.2}$$

For each $t \in [0,1]$ and each $n \geq 1$, let $\delta_n(t) = t_k^n$, $\theta_n(t) = t_{k+1}^n$, if $t \in]t_k^n, t_{k+1}^n]$ and $\delta_n(0) = \theta_n(0) = 0$. So by (2.4.2) we get

$$\dot{u}_n(t) \in -A(\theta_n(t))u_n(\theta_n(t)) + \sigma(\delta_n(t), \tau(\delta_n(t))u_n), \tag{2.4.3}$$

for a.e $t \in [0,1]$. It is obvious that, for all $n \geq 1$ and for all $t \in [0,1]$, the following hold:

$$\sigma(\delta_n(t), \tau(\delta_n(t))u_n) \in F(\delta_n(t), \tau(\delta_n(t))u_n), \tag{2.4.4}$$

$$u_n(\delta_n(t)), u_n(\theta_n(t)) \in \overline{D}, \tag{2.4.5}$$

$$\lim_{n\to\infty} \delta_n(t) = \lim_{n\to\infty} \theta_n(t) = t. \tag{2.4.6}$$

Claim: $(\dot{u}_n)$ *is bounded a.e.*

Pick $s > 0$ such that $||\varphi(0)|| < s < +\infty$. By (H_2) there is $\delta(s)$ such that

$$||J_{e_n}A(t_0^n)\varphi(0) - \varphi(0)|| \leq e_n\delta(s).$$

Let ρ as in (H_1). There is $\gamma > 0$ such that $||\rho(t)|| \leq \frac{\gamma}{2}$ for all $t \in [0,1]$. By using (H_1) and the preceding inequality, we obtain the estimate

$$\begin{aligned}
||x_k^n - \varphi(0)|| &\leq ||J_{e_n}A(t_k^n)(x_{k-1}^n + e_n\sigma(t_{k-1}^n, \tau(t_{k-1}^n)u_n)) - J_{e_n}A(t_k^n)\varphi(0)|| \\
&\quad + ||J_{e_n}A(t_k^n)\varphi(0) - J_{e_n}A(t_0^n)\varphi(0)|| \\
&\quad + ||J_{e_n}A(t_0^n)\varphi(0) - \varphi(0)|| \\
&\leq ||x_{k-1}^n - \varphi(0)|| + e_n||\sigma(t_{k-1}^n, \tau(t_{k-1}^n)u_n)|| \\
&\quad + e_n\gamma L(||\varphi(0)||) + e_n\delta(s) \\
&\leq ||x_{k-1}^n - \varphi(0)|| + e_n(|K| + \gamma L(||\varphi(0)||) + \delta(s)).
\end{aligned}$$

Iterating the preceding inequality gives:

$$||x_k^n - \varphi(0)|| \leq t_k^n(|K| + \gamma L(||\varphi(0)||) + \delta(s)), \tag{2.4.7}$$

for all $n \geq 1$ and for all $k = 1, 2, ..., n$. So we have

$$||x_k^n|| \leq \beta := ||\varphi(0)|| + |K| + \gamma L(||\varphi(0)||) + \delta(s), \tag{2.4.8}$$

for all $n \geq 1$ and for all $k = 0, 1, 2, .., n$. Using (2.4.8) and (H_2), there is $\zeta(\beta) > 0$ such that $||J_{e_n}A(t_0^n)x_k^n - x_k^n|| \leq e_n\zeta(\beta)$. So by (H_1) and (H_2), we get the estimate

$$\begin{aligned}
||x_{k+1}^n - x_k^n|| &\leq ||J_{e_n}A(t_{k+1}^n)(x_k^n + e_n\sigma(t_k^n, \tau(t_k^n)u_n)) - J_{e_n}A(t_{k+1}^n)x_k^n|| \\
&\quad + ||J_{e_n}A(t_{k+1}^n)x_k^n - J_{e_n}A(t_0^n)x_k^n|| \\
&\quad + ||J_{e_n}A(t_0^n)x_k^n - x_k^n|| \\
&\leq e_n(|K| + \gamma L(\beta) + \zeta(\beta))
\end{aligned}$$

for all $n \geq 1$ and for all $k = 0, 1, 2, .., n$. By (2.4.2) and the preceding estimate, we get

$$||\dot{u}_n(t)|| \leq m := |K| + \gamma L(\beta) + \zeta(\beta), \tag{2.4.9}$$

for all $n \geq 1$ and a.e $t \in [0, 1]$. That proves the claim. By (2.4.6) and (2.4.9), we have

$$\lim_{n\to\infty} ||u_n(\theta_n(t)) - u_n(t)|| = 0, \tag{2.4.10}$$

for all $t \in [0, 1]$. By (2.4.8) $(u_n(\theta_n(t))) \in \overline{D} \cap \overline{B}_E(0, \beta)$ for all $t \in [0, 1]$ and for all $n \in \mathbb{N}$. As $\overline{D}$ is ball-compact, $(u_n(\theta_n(t)))$ is relatively compact in E for every $t \in [0, 1]$, so is $(u_n(t))$. Thus $(u_n(.))$ is relatively compact in $\mathcal{C}_E([-r, 1])$. Hence we may suppose that $(\dot{u}_n)$ weakly converges in $L^2_E([0, 1])$ to a function v with $||v(t)|| \leq m$ for a.e $t \in [0, 1]$, and (u_n) converges in $\mathcal{C}_E([-r, 1])$ to a function u with

$$u(t) = \varphi(0) + \int_0^t v(s)\, ds, \ \ \forall t \in [0, 1].$$

As $||\sigma(\delta_n(t), \tau(\delta_n(t))u_n)|| \leq |K|$ for all $n \geq 1$ and for all $t \in [0, 1]$, we may suppose that $(g_n(.)) := (\sigma(\delta_n(.), \tau(\delta_n(.))u_n))$ converges weakly in $L^2_E([0, 1])$ to a function g with $||g(t)|| \leq |K|$ for a.e $t \in [0, 1]$.

Claim: $v(t) \in -A(t)u(t) + g(t)$ *a.e* $t \in [0, 1]$.

Let us set $w_n(t) := J_{e_n}A(t)(u_n(\delta_n(t)) + e_n g_n(t))$ for all $n \geq 1$ and for all $t \in [0, 1]$. Then $w_n(t) \in D(A(t))$ for all $t \in [0, 1]$ and we have

$$u_n(\delta_n(t)) + e_n g_n(t) \in w_n(t) + e_n A(t) w_n(t). \tag{2.4.11}$$

In view of (H_1) and (2.4.8) we have the estimate

$$\begin{aligned} ||u_n(\theta_n(t)) - w_n(t)|| &\leq e_n|\rho(\theta_n(t)) - \rho(t)|L(||u_n(\delta_n(t)) + e_n g_n(t)||) \\ &\leq e_n|\rho(\theta_n(t)) - \rho(t)|L(\beta + |K|). \end{aligned}$$

It follows that

$$\lim_{n\to\infty} ||u_n(\theta_n(t)) - w_n(t)|| = 0, \ \ \forall t \in [0, 1]. \tag{2.4.12}$$

Consequently we get

$$\lim_{n\to\infty} w_n(t) = u(t), \ \ \forall t \in [0, 1]. \tag{2.4.13}$$

Let $n \geq 1$ and let $t \in]0, 1[$. Then $t \in]t^n_k, t^n_{k+1}[$ for some $0 \leq k < 1$. So, taking (2.4.11) and the preceding estimate into account, we get

$$
\begin{aligned}
&d(-\dot{u}_n(t) + g_n(t), A(t)w_n(t)) \\
&= d\left(\frac{u_n(\delta_n(t)) - u_n(\theta_n(t))}{e_n} + g_n(t), A(t)w_n(t)\right) \\
&\leq \frac{1}{e_n}||u_n(\theta_n(t)) - w_n(t)|| \\
&\leq |\rho(\theta_n(t)) - \rho(t)|L(\beta + |K|).
\end{aligned}
$$

Since $A(t)w_n(t)$ is closed and convex, the preceding inequality implies that

$$
-\dot{u}_n(t) + g_n(t) \in A(t)w_n(t) + r_n(t)\overline{B}_E
$$

a.e with $r_n(t) := |\rho(\theta_n(t)) - \rho(t)|L(\beta + |K|) \to 0$ for all $t \in [0,1]$. As $-\dot{u}_n(.) + g_n(.) \to -v + g$ weakly in $L^2_E([0,1])$ and $w_n \to u$ strongly in $L^2_E([0,1])$ by (2.4.8), (2.4.12) and (2.4.13), from Lemma 2.3, we deduce that

$$
v(t) \in -A(t)u(t) + g(t) \;\; a.e \; t \in [0,1].
$$

Claim: $g(t) \in F(t, \tau(t)u)$ $a.e$ $t \in [0,1]$. Let $t \in [0,1]$. Using (2.4.9) we get

$$
\begin{aligned}
&||\tau(\delta_n(t))u_n - \tau(t)u||_{C_E([-r,0])} \\
&\leq ||\tau(\delta_n(t))u_n - \tau(t)u_n||_{C_E([-r,0])} + ||\tau(t)u_n - \tau(t)u||_{C_E([-r,0])} \\
&\leq \sup_{\{s_1,s_2\in[-r,1],|s_1-s_2|<e_n\}} ||u_n(s_1) - u_n(s_2)|| + ||\tau(t)u_n - \tau(t)u||_{C_E([-r,0])} \\
&\leq \sup_{\{s_1,s_2\in[-r,0],|s_1-s_2|<e_n\}} ||u_n(s_1) - u_n(s_2)|| \\
&\quad + \sup_{\{s_1,s_2\in[0,1],|s_1-s_2|<e_n\}} ||u_n(s_1) - u_n(s_2)|| \\
&\quad + ||\tau(t)u_n - \tau(t)u||_{C_E([-r,0])} \\
&\leq \sup_{\{s_1,s_2\in[-r,0],|s_1-s_2|<e_n\}} ||\varphi(s_1) - \varphi(s_2)|| \\
&\quad + me_n + ||\tau(t)u_n - \tau(t)u||_{C_E([-r,0])}.
\end{aligned}
$$

Using the continuity of φ, the uniform convergence of u_n towards u, and the preceding estimate, we see that

$$
\lim_{n\to\infty} ||\tau(\delta_n(t))u_n - \tau(t)u||_{C_E([-r,0])} = 0.
$$

As $g_n(t) := \sigma(\delta_n(t), \tau(\delta_n(t))u_n) \in F(\delta_n(t), \tau(\delta_n(t))u_n)$ for all $n \geq 1$ and for all $t \in [0,1]$, $\delta_n(t) \to t$ in $[0,1]$ and, $\tau(\delta_n(t))u_n \to \tau(t)u$ in $C_E([-r,0])$ pointwisely on $[0,1]$, by invoking the scalarly upper semicontinuity of F on $[0,1] \times C_E([-r,0])$ and the closure result in ([9], Theorem V-14), we conclude that $g(t) \in F(t, \tau(t)u)$ a.e. Thus we have completely proved the existence of strong solutions of (P_τ) in the case when F is globally scalarly upper semicontinuous on $[0,1] \times \mathcal{C}_E([-r,0])$.

Step 2. Now let us go into the case when F satisfies (H_4). Let us summarize the results obtained in the first step. Given $\varphi \in \mathcal{C}_E([-r,0])$ with $\varphi(0) \in \overline{D}$, there is continuous mapping $u \in \mathcal{C}_E([-r,1])$ such that its restriction to $[0,1]$ is an absolutely continuous solution of the evolution equation

$$\dot{u}(t) \in -A(t)u(t) + F(t,\tau(t)u) \ a.e\ t \in [0,1],$$

and $u(s) = \varphi(s)$, $\forall s \in [-r,0]$. Moreover we have $u(t) = \varphi(0) + \int_0^t \dot{u}\, ds$, $\forall t \in [0,1]$ with $u(t) \in \bar{D}, \forall t \in [0,1]$ and $||\dot{u}(t)|| \leq m$ a.e $t \in [0,1]$ where m is given by the estimate (2.4.9). Let K_σ denotes K equipped with the $\sigma(E,E')$ topology. Let us denote by $ck(K_\sigma)$ the set of all nonempty convex compact subsets of K_σ. By a recent version of multivalued Scorza-Dragoni theorem [6], there exists a multifunction $F_0 : [0,1] \times \mathcal{C}_E([-r,0]) \to ck(K_\sigma) \cup \{\emptyset\}$ which enjoys the following properties :

(1) there is a Lebesgue negligible set $N \subset [0,1]$ independent of (t,u) such that $F_0(t,u) \subset F(t,u)$, $\forall t \in [0,1] \setminus N$, $\forall u \in \mathcal{C}_E([-r,0])$,

(2) if $x : [0,1] \to \mathcal{C}_E([-r,0])$ and $y : [0,1] \to E$ are Lebesgue-measurable functions with $y(t) \in F(t,x(t))$ a.e, then $y(t) \in F_0(t,x(t))$ a.e,

(3) for every $\varepsilon > 0$, there is a compact subset $J_\varepsilon \subset [0,1]$ such that the Lebesgue measure of $[0,1] \setminus J_\varepsilon$ is less that ε, and the graph of the restriction $F_0|J_\varepsilon \times \mathcal{C}_E([-r,0])$ is closed in $J_\varepsilon \times \mathcal{C}_E([-r,0]) \times K_\sigma$, and $\{\emptyset\} \neq F_0(t,u) \subset F(t,u)$, $\forall (t,u) \in J_\varepsilon \times \mathcal{C}_E([-r,0])$.

By (3) there exists an increasing sequence of compact sets (J_n) in $[0,1]$ such that the Lebesgue measure of $[0,1] \setminus J_n$ tends to 0 when $n \to \infty$ and that the graph of the restriction of F_0 on $J_n \times \mathcal{C}_E([-r,0])$ is closed in $J_n \times \mathcal{C}_E([-r,0]) \times K_\sigma$ (i.e, it is upper semicontinuous) and has nonempty values.

Let $\widetilde{F}_n$ be a $ck(K_\sigma)$-valued upper semicontinuous extension (see [2]) of $F_0|J_n \times \mathcal{C}_E([-r,0])$ to $[0,1] \times \mathcal{C}_E([-r,0])$, with $\widetilde{F}_n(t,u) \subset K_\sigma$, for all $(t,u) \in [0,1] \times \mathcal{C}_E([-r,0])$.

We now apply step 1 with $\widetilde{F}_n$ substituted for F. Thus, using the above remark, for every n, there is a continuous mapping $u_n \in \mathcal{C}_E([-r,1])$ such that its restriction to $[0,1]$ is an absolutely continuous solution of the evolution equation

$$\dot{u}_n(t) \in -A(t)u_n(t) + \widetilde{F}_n(t,\tau(t)u_n) \ a.e\ t \in [0,1],$$

with $u_n(s) = \varphi(s)$, $\forall s \in [-r,0]$ and $u_n(t) = \varphi(0) + \int_0^t \dot{u}_n(s)\, ds$, $\forall t \in [0,1]$ with $u_n(t) \in \bar{D}, \forall t \in [0,1]$, and $||\dot{u}_n(t)|| \leq m$, a.e $t \in [0,1]$. As $\bar{D}$ is ball compact, we may suppose that (u_n) converges uniformly to $u \in$

$\mathcal{C}_E([-r,1])$, $(\dot{u}_n)$ weakly* converges in $L^\infty_E([0,1])$ to $\dot{u}$ with $||\dot{u}(t)|| \le m$ a.e, and $u(t) = \varphi(0) + \int_0^t \dot{u}(s)\,ds,\ \forall t \in [0,1]$. We now finish the proof as follows. There is a measurable mapping $z_n : [0,1] \to K$ with $z_n(t) \in \widetilde{F}_n(t, \tau(t)u_n)$ for a.e $t \in [0,1]$ and

$$\dot{u}_n(t) \in -A(t)u_n(t) + z_n(t),$$

a.e $t \in [0,1]$. We may suppose that $z_n \to z$ weakly* in $L^\infty_E([0,1])$. By construction, there is a Lebesgue null set N_n such that $z_n(t) \in F_0(t, \tau(t)u_n)$ for all $t \in J_n \setminus N_n$. Let $N_0 := ([0,1] \setminus \cup_n J_n) \cup (\cup_n N_n)$ which is Lebesgue-negligible. If $t \notin N_0$, there is an integer $p := p(t)$ such that $z_n(t) \in F_0(t, \tau(t)u_n)$ for all $n \ge p(t)$. As F_0 is scalarly upper semicontinuous on J_p and $\tau(t)u_n \to \tau(t)u$ in $\mathcal{C}_E([-r,0])$, we have

$$\limsup_n \langle x', z_n(t)\rangle \le \limsup_n \delta^*(x', F_0(t, \tau(t)u_n)) \le \delta^*(x', F_0(t, \tau(t)u)),$$

for all $x' \in E'$ and for all $n \ge p$. Thus

$$\limsup_n \langle x', z_n(t)\rangle \le \delta^*(x', F_0(t, \tau(t)u),$$

for a.e $t \in [0,1]$. It follows that, for every measurable set $A \in [0,1]$ and for every $x' \in E'$,

$$\lim_n \int_A \langle x', z_n(t)\rangle\,dt = \int_A \langle x', z(t)\rangle\,dt \le \int_A \delta^*(x', F_0(t, \tau(t)u)\,dt,$$

using Fatou's lemma. Consequently $z(t) \in F_0(t, \tau(t)u) \subset F(t, \tau(t)u)$ a.e.

□

We have immediately the following corollaries which are new and will be used later.

Corollary 2.5. *Suppose that $A(t) : E \to cc(E) \cup \{\emptyset\}; t \in [0,1]$, is an m-accretive operator, $(H_1), (H_2), (H_3)$ are satisfied, and $\overline{D}$ is ball-compact, K is a convex weakly compact subset of E and $f : [0,1] \times \mathcal{C}_E([-r,0]) \to K$ is a Carathéodory mapping, that is, f is separately measurable on $[0,1]$ and separately continuous on $\mathcal{C}_E([-r,0])$, then for any $\varphi \in \mathcal{C}_E([-r,0])$ with $\varphi(0) \in \overline{D}$, there is a continuous mapping $u \in \mathcal{C}_E([-r,1])$ satisfying*

$$(P_\tau)\begin{cases} \dot{u}(t) \in -A(t)u(t) + f(t, \tau(t)u)\ a.e\ t \in [0,1], \\ u(s) = \varphi(s),\ \forall s \in [-r,0];\ u(t) \in \overline{D},\ \forall t \in [0,1]. \end{cases}$$

Another immediate corollary of Theorem 2.4 concerns an undelayed differential inclusion governed by an m-accretive operator of the form

$$(P_0)\begin{cases}\dot{u}(t) \in -A(t)u(t) + G(t, u(t)) \ a.e\ t \in [0, 1],\\ u(0) = u_0 \in \overline{D},\end{cases}$$

where G is a convex weakly compact valued perturbation defined on $[0, 1] \times E$.

Corollary 2.6. *Suppose that* $A(t) : E \to cc(E) \cup \{\emptyset\}; t \in [0, 1]$, *is an m-accretive operator,* $(H_1), (H_2), (H_3))$ *are satisfied, and* $\overline{D}$ *is ball-compact,* K *is a convex weakly compact subset of* E *and* $G : [0, 1] \times E \to K$ *is a separately scalarly measurable on* $[0, 1]$ *and separately scalarly upper semicontinuous on* E*, then for any* $u_0 \in \overline{D}$*, there is an absolutely continuous solution* $u : [0, 1] \to E$ *to*

$$(P_0)\begin{cases}\dot{u}(t) \in -A(t)u(t) + G(t, u(t)) \ a.e\ t \in [0, 1],\\ u(0) = u_0 \in \overline{D}.\end{cases}$$

For single valued perturbation $G(t, x) = \{g(t, x)\}$, Corollary 2.6 is reduced to the following

Corollary 2.7. *Suppose that* $A(t) : E \to cc(E) \cup \{\emptyset\}; t \in [0, 1]$, *is an m-accretive operator,* $(H_1), (H_2), (H_3)$ *are satisfied, and* $\overline{D}$ *is ball-compact,* K *is a convex weakly compact subset of* E *and* $g : [0, 1] \times E \to K$ *is a Carathéodory mapping, that is,* g *is separately measurable on* $[0, 1]$ *and separately continuous on* E*, then for any* $u_0 \in \overline{D}$*, there is an absolutely continuous solution* $u : [0, 1] \to E$ *to*

$$(P_0)\begin{cases}\dot{u}(t) \in -A(t)u(t) + g(t, u(t)) \ a.e\ t \in [0, 1],\\ u(0) = u_0 \in \overline{D}.\end{cases}$$

The following is a viability result illustrating the techniques developed in the proof of Theorem 2.4. See [19] for the existence of BV solutions for an evolution equation governed by a maximal monotone operator $A(t)$ with time-dependant domain in Hilbert space.

Proposition 2.8. *Suppose that* $A(t) : E \to cc(E) \cup \{\emptyset\}; t \in [0, 1]$, *is an m-accretive operator satisfying* $(H_1), (H_2), (H_3)$ *and* $C : [0, 1] \to E$ *is a closed ball-compact valued, (that is,* $\forall t \in [0, 1], C(t)$ *is closed and its intersection with any closed ball of* E *is compact) upper semicontinuous multifunction satisfying: (*)* $0 \in C(t)$*, for all* $t \in [0, 1]$ *and (**)* $[I_E + \lambda A(t)]^{-1}C(s) \subset C(t)$*, for all* $\lambda > 0$ *and for all* $0 \le s \le t \le 1$*. Then for any* $a \in C(0) \cap \overline{D}$*, there is an absolutely continuous solution* $u : [0, 1] \to E$ *to*

$$\begin{cases}\dot{u}(t) \in -A(t)u(t) \ a.e\ t \in [0, 1],\\ u(0) = a, \ \ u(t) \in C(t) \cap \overline{D}, \ \ \forall t \in [0, 1].\end{cases}$$

Proof. That is an easy adaptation of the proof of Theorem 2.4, so we do not want to go into details too much. We will follow the discretization technique developed in Theorem 2.4. By (**) we first observe that $[I_E+\lambda A(t)]^{-1}C(s) \subset C(t) \cap \bar{D}$, for all $\lambda > 0$ and for all $0 \leq s \leq t \leq 1$. We define a sequence of absolutely continuous functions (u_n) in $C_E([0,1])$ such that a subsequence of its converges uniformly to an absolutely continuous solution of the equation under consideration. Let $n \geq 1$ be a fixed integer. In order to construct u_n on $[0,1]$, we consider a partition of $[0,1]$ by the points $t_k^n = ke_n$, $e_n = \frac{1}{n}, k = 0,1,2,...,n$. We define u_n by linear interpolation, where $u_n(t_k^n) = x_k^n \in C(t_k^n) \cap \overline{D(A(t_k^n))}$, $k = 0,1,2,...,n$ are obtained by induction starting with $u_n(0) = x_0^n = a \in C(0) \cap \overline{D(0)}$. Let us set

$$x_1^n := J_{e_n}A(t_1^n)x_0^n.$$

We have $x_1^n \in C(t_1^n) \cap D(A(t_1^n)) \subset C(t_1^n) \cap \overline{D}$. For $t \in]t_0^n, t_1^n]$, let us set

$$u_n(t) = \frac{t_1^n - t}{e_n}x_0^n + \frac{t - t_0^n}{e_n}x_1^n.$$

Then for $t \in]t_0^n, t_1^n[$, we have

$$\dot{u}_n(t) = \frac{x_1^n - x_0^n}{e_n} \in -A(t_1^n)x_1^n.$$

By induction on $0 \leq k \leq n$ we set

$$x_{k+1}^n := J_{e_n}A(t_{k+1}^n)x_k^n. \tag{2.8.1}$$

Then $x_{k+1}^n \in C(t_{k+1}^n) \cap D(A(t_{k+1}^n)) \subset C(t_{k+1}^n) \cap \overline{D}$ for $k = 0,1,2,...,n$. For $]t_k^n, t_{k+1}^n]$, $0 \leq k \leq n-1$, we set

$$u_n(t) = \frac{t_{k+1}^n - t}{e_n}x_k^n + \frac{t - t_k^n}{e_n}x_{k+1}^n,$$

then, for $]t_k^n, t_{k+1}^n[$, we have

$$\dot{u}_n(t) = \frac{x_{k+1}^n - x_k^n}{e_n} \in -A(t_{k+1}^n)x_{k+1}^n. \tag{2.8.2}$$

For each $t \in [0,1]$ and each $n \geq 1$, let $\delta_n(t) = t_k^n$, $\theta_n(t) = t_{k+1}^n$, if $t \in]t_k^n, t_{k+1}^n]$ and $\delta_n(0) = \theta_n(0) = 0$. So by (2.8.2) we get

$$\dot{u}_n(t) \in -A(\theta_n(t))u_n(\theta_n(t)), \tag{2.8.3}$$

for a.e $t \in [0,1]$. It is obvious that, for all $n \geq 1$ and for all $t \in [0,1]$, the following hold:

$$u_n(\delta_n(t)) \in C(\delta_n(t)) \cap \overline{D},\ u_n(\theta_n(t)) \in C(\theta_n(t)) \cap \overline{D}, \tag{2.8.4}$$

(2.8.5) $$\lim_{n\to\infty} \delta_n(t) = \lim_{n\to\infty} \theta_n(t) = t.$$

Claim: $(\dot{u}_n)$ *is bounded a.e.*

Pick $s > 0$ such that $||a|| < s < +\infty$. Since the operator $A(t_0^n)$ satisfies (H_2), there is $\delta(s)$ such that

$$||J_{e_n} A(t_0^n)a - a|| \leq e_n \delta(s).$$

Let ρ as in (H_1). There is $\gamma > 0$ such that $||\rho(t)|| \leq \frac{\gamma}{2}$ for all $t \in [0,1]$. By using (H_1) and the preceding inequality, we obtain the estimate

$$\begin{aligned} ||x_k^n - a|| &\leq ||J_{e_n} A(t_k^n)x_{k-1}^n - J_{e_n} A(t_k^n)a|| \\ &\quad + ||J_{e_n} A(t_k^n)a - J_{e_n} A(t_0^n)a|| \\ &\quad + ||J_{e_n} A(t_0^n)a - a|| \\ &\leq ||x_{k-1}^n - a|| + e_n \gamma L(||a||) + e_n \delta(s) \\ &= ||x_{k-1}^n - a|| + e_n(\gamma L(||a||) + \delta(s)). \end{aligned}$$

Iterating the preceding inequality gives

(2.8.6) $$||x_k^n - a|| \leq t_k^n(\gamma L(||a||) + \delta(s)),$$

for all $n \geq 1$ and for all $k = 1, 2, ..., n$. So we have

(2.8.7) $$||x_k^n|| \leq \beta := ||a|| + \gamma L(||a||) + \delta(s),$$

for all $n \geq 1$ and for all $k = 0, 1, 2, .., n$. Using (2.8.7) and (H_2) there is $\zeta(\beta) > 0$ such that $||J_{e_n} A(t_0^n)x_k^n - x_k^n|| \leq e_n \zeta(\beta)$. So, by (H_1) we get the estimate

$$\begin{aligned} ||x_{k+1}^n - x_k^n|| &\leq ||J_{e_n} A(t_{k+1}^n)x_k^n - J_{e_n} A(t_0^n)x_k^n|| + ||J_{e_n} A(t_0^n)x_k^n - x_k^n|| \\ &\leq e_n(\gamma L(\beta) + \zeta(\beta)) \end{aligned}$$

for all $n \geq 1$ and for all $k = 0, 1, 2, .., n$. By (2.8.2) and the preceding estimate, we get

(2.8.8) $$||\dot{u}_n(t)|| \leq m := \gamma L(\beta) + \zeta(\beta),$$

for all $n \geq 1$ and a.e $t \in [0,1]$. Hence the claim is proved. It is obvious that (u_n) is equicontinuous. By (2.8.5) and (2.8.8), we have

(2.8.9) $$\lim_{n\to\infty} ||u_n(\theta_n(t)) - u_n(t)|| = 0,$$

for all $t \in [0,1]$. By construction and (2.8.7), we have $x_k^n \in C(t_k^n) \cap \beta \overline{B}_H$, for all $n \geq 1$ and all $k = 0, 1, 2, ..., n$. So, for every $t \in [0,1]$, we have

$u_n(\theta_n(t)) \in C(\theta_n(t)) \cap \beta\overline{B}_H$. In view of (*), upper semicontinuity of $C(.)$ and ball-compactness assumption on the closed sets $C(t)$, we see that the nonempty compact valued multifunction $C(.) \cap \beta\overline{B}_H$ is upper semicontinuous on $[0,1]$. Thus, by (2.8.5) we conclude that the sequence $(u_n(\theta_n(t))$ is relatively compact in H. So, by Ascoli's theorem, the sequence (u_n) is relatively compact in $\mathcal{C}_H([0,1])$. It is obvious by using (2.8.8), that the sequence $(\dot{u}_n)$ is relatively weakly compact in $L^2_H([0,1])$. Hence we may suppose that $(\dot{u}_n)$ weakly converges in $L^2_H([0,1])$ to a function v with $||v(t)|| \leq m$ for a.e $t \in [0,1]$, and (u_n) converges in $\mathcal{C}_H([0,1])$ to a function u with

$$u(t) = a + \int_0^t v(s)\,ds, \quad \forall t \in [0,1].$$

Claim: $v(t) \in -A(t)u(t)$ *a.e* $t \in [0,1]$ *and conclusion.*

Let us set $w_n(t) := J_{e_n}A(t)(u_n(\delta_n(t))$ for all $n \geq 1$ and for all $t \in [0,1]$. Then we have

$$u_n(\delta_n(t)) \in w_n(t) + e_n A(t) w_n(t). \tag{2.8.10}$$

In view of (H_1) and (2.8.7) we have the estimate

$$\begin{aligned} ||u_n(\theta_n(t)) - w_n(t)|| &\leq e_n|\rho(\theta_n(t)) - \rho(t)|L(||u_n(\delta_n(t))||) \\ &\leq e_n|\rho(\theta_n(t)) - \rho(t)|L(\beta). \end{aligned}$$

It follows that

$$\lim_{n\to\infty} ||u_n(\theta_n(t)) - w_n(t)|| = 0, \ \ \forall t \in [0,1]. \tag{2.8.11}$$

Consequently we get

$$\lim_{n\to\infty} w_n(t) = u(t), \ \ \forall t \in [0,1]. \tag{2.8.12}$$

Let $n \geq 1$ and let $t \in]0,1[$. Then $t \in]t^n_k, t^n_{k+1}[$ for some $0 \leq k < 1$. So, by taking (2.8.10) and the preceding estimate into account we get

$$\begin{aligned} d(-\dot{u}_n(t), A(t)w_n(t)) &= d\left(\frac{u_n(\delta_n(t)) - u_n(\theta_n(t))}{e_n}, A(t)w_n(t)\right) \\ &\leq \frac{1}{e_n}||u_n(\theta_n(t)) - w_n(t)|| \\ &\leq |\rho(\theta_n(t)) - \rho(t)|L(\beta). \end{aligned}$$

As $||\rho(\theta_n(t)) - \rho(t)||L(\beta) \to 0$ for all $t \in [0,1]$, $\dot{u}_n(.) \to v$ weakly in $L^2_H([0,1])$ and from (2.8.7)-(2.8.11)-(2.8.12) and Lebesgue's theorem, $w_n \to u$ strongly in $L^2_H([0,1])$, by using Lemma 2.3, we get

$$v(t) \in -A(t)u(t) \ a.e\ t \in [0,1].$$

As the graph of the compact valued upper semicontinuous multifunction $C(.)\cap \beta\overline{B}_H$ is closed, from (2.8.9) and our construction we see that $u(t) \in C(t) \cap \beta\overline{B}_H \cap \overline{D}$. The proof is therefore complete. □

The following is a uniqueness result whose proof is borrowed from ([7], Proposition 2.2).

Proposition 2.9. *Suppose that* $A(t) : E \to 2^E; t \in [0,1]$, *is an* m*-accretive operator. Let* $g : [0,1] \times C_E([-r,0]) \to E$ *such that:*

(i) there exits $d > 0$ *such that* $||g(t,u) - g(t,v)|| \leq d||u - v||_{C_E([-r,0])}$ *for all* $t \in [0,1]$ *and for all* $u, v \in C_E([-r,0])$,

(ii) for every $u \in C_E([-r,0])$, *the function* $g(.,u)$ *is integrable on* $[0,1]$.

Let $\psi \in C_E([-r,0])$ *with* $\psi(0) \in \bar{D}$. *If* u_1 *and* u_2 *are strong solutions for the problem*

$$\begin{cases} \dot{u}(t) \in -A(t)u(t) + g(t, \tau(t)u), \ \ a.e\ t \in [0,1], \\ u(s) = \psi(s), \forall s \in [-r,0], \end{cases}$$

then $||u_1 - u_2||_t = 0, \ \ \forall t \in [0,1]$ *where* $||.||_t$ *denotes the norm of* $C_E([-r,t])$.

Proof. Let u_1, u_2 be two strong solutions of the problem under consideration. Then we have $g(t,\tau(t)u_1) - \dot{u}_1(t) \in A(t)u_1(t)$ and $g(t,\tau(t)u_2) - \dot{u}_2(t) \in A(t)u_2(t)$ a.e. Since $A(t)$ is accretive, then

$$(2.9.1)\quad \langle u_1(t) - u_2(t), g(t,\tau(t)u_1) - \dot{u}_1(t) - g(t,\tau(t)u_2) + \dot{u}_2(t)\rangle_+ \geq 0,$$

a.e, where $\langle .,.\rangle_+$ denotes the usual upper-semi-inner product on E. It is well known (see e.g [31]) that (2.9.1) can be written as

$$(2.9.2)\quad \begin{aligned} 0 &\leq \langle u_1(t) - u_2(t), g(t,\tau(t)u_1) - \dot{u}_1(t) - g(t,\tau(t)u_2) + \dot{u}_2(t)\rangle_+ \\ &= \langle j(u_1(t) - u_2(t)), g(t,\tau(t)u_1) - \dot{u}_1(t) - g(t,\tau(t)u_2) + \dot{u}_2(t)\rangle, \end{aligned}$$

where $j : E \to E'$ is the single-valued duality mapping. So (2.9.2) is equivalent to

$$\langle j(u_1(t) - u_2(t)), \dot{u}_1(t) - \dot{u}_2(t)\rangle \leq \langle j(u_1(t) - u_2(t)), g(t,\tau(t)u_1) - g(t,\tau(t)u_2)\rangle.$$

By ([31], Lemma 1.4.2(ix)) we have

$$\begin{aligned} \frac{1}{2}\frac{d}{dt}||u_1(t) - u_2(t)||^2 &\leq \langle j(u_1(t) - u_2(t)), g(t,\tau(t)u_1) - g(t,\tau(t)u_2)\rangle \\ &\leq d\,||u_1(t) - u_2(t)||\,||\tau(t)u_1 - \tau(t)u_2||_{C_E([-r,0])}, \end{aligned}$$

for a.e $t \in [0,1]$. Integrating on $[0,s]$ we obtain

$$\frac{1}{2}||u_1(s)-u_2(s)||^2 \le d\int_0^t ||\tau(\theta)u_1-\tau(\theta)u_2||_{C_E([-r,0])}\,||u_1(\theta)-u_2(\theta)||\,d\theta,$$
$$0 \le s \le t.$$

Since $u_1 = \psi = u_2$ in $[-r,0]$, the last inequality implies that

$$\frac{1}{2}||u_1-u_2||_t^2 \le d\int_0^t ||u_1-u_2||_\theta^2\,d\theta,$$

where $||.||_\theta$ denotes the norm of $C_E([-r,\theta])$. Because $\theta \mapsto ||u_1-u_2||_\theta$ is continuous, by applying Gronwall's lemma we conclude that

$$||u_1-u_2||_t = 0, \ \ \forall t \in [0,1].$$

Hence $u_1 = u_2$ in $[-r,1]$. $\square$

The following deals with compactness of solutions set for problems (P_0). For any $cc(E)$-valued measurable multifunction $\Gamma : [0,1] \to cc(E)$, let us denote by $\mathcal{S}_\Gamma^2 := \{f \in L_E^2([0,1] : f(t) \in \Gamma(t)\ a.e\}$.

Proposition 2.10. *Suppose that $A(t) : E \to cc(E) \cup \{\emptyset\}; t \in [0,1]$, is an m-accretive operator, $(H_1), (H_2), (H_3)$ are satisfied, and $\overline{D}$ is ball-compact. Then the following hold:*

a) For any $x_0 \in \overline{D}$ and for any convex weakly compact subset Γ, the strong solutions set $\{u_f : f \in \mathcal{S}_\Gamma^2\}$ of the problem

$$(P_0)\begin{cases}\dot{u}(t) \in -A(t)u(t) + f(t)\ a.e\ t \in [0,1],\\ u(0) = x_0,\ u(t) \in \overline{D},\ \forall t \in [0,1],\end{cases}$$

is a nonempty compact subset of $C_E([0,1])$.

b) For any $x_0 \in \overline{D}$ and for any convex weakly compact subset Γ, for any convex weakly compact valued multifunction $G : [0,1] \times E \to \Gamma$, separately scalarly measurable and separately scalarly upper semicontinuous, the strong solutions set of

$$(P_0)\begin{cases}\dot{u}(t) \in -A(t)u(t) + G(t,u(t))\ a.e\ t \in [0,1],\\ u(0) = x_0,\ u(t) \in \overline{D},\ \forall t \in [0,1],\end{cases}$$

is a nonempty compact subset of $C_E([0,1])$.

Proof. a) An easy inspection of Step 1 of the proof of Theorem 2.4 shows that the strong solutions set $\{u_f : f \in \mathcal{S}_\Gamma^2\}$ of

$$(P_0)\begin{cases}\dot{u}(t) \in -A(t)u(t) + f(t) \ a.e\ t \in [0,1],\\ u(0) = x_0,\ u(t) \in \overline{D},\ \forall t \in [0,1],\end{cases}$$

is not empty. Moreover, we have $u_f(t) = x_0 + \int_0^t \dot{u}_f(s)\,ds \in \overline{D}$, $\forall f \in \mathcal{S}^2_\Gamma$, $\forall t \in [0,1]$ with $||\dot{u}_f(t)|| \leq M$ for all $f \in \mathcal{S}^2_\Gamma$ and for all $t \in [0,1]$, where M is a positive constant, using the estimate (2.4.9) for this particular case. By Ascoli's theorem, $\{u_f : f \in \mathcal{S}^2_\Gamma\}$ is relatively compact in $\mathcal{C}_E([0,1])$. Let (f_n) be a sequence in $\mathcal{S}^2_\Gamma$. There is a subsequence (g_m) of (f_n) such that (g_m) converges $\sigma(L^2_E([0,1]), L^2_{E'}([0,1]))$ to $h \in \mathcal{S}^2_\Gamma$. We may also suppose that $(\dot{u}_{g_m})$ converges $\sigma(L^2_E([0,1]), L^2_{E'}([0,1]))$ to a measurable function v with $||v(t)|| \leq M$ a.e and (u_{g_m}) converges to $u \in \mathcal{C}_E([0,1])$ with $u(t) = x_0 + \int_0^t \dot{u}(s)\,ds$ for all $t \in [0,1]$, and $\dot{u} = v$ a.e. By Lemma 2.3 we see that $\dot{u}(t) \in -A(t)u(t) + h(t)$ for a.e $t \in [0,1]$.

b) It is easy to see that the strong solutions set to problem (P_0) is relatively compact in $\mathcal{C}_E([0,1])$, using a). Let (u_n) be a sequence of strong solutions to (P_0) converging uniformly to a function $u_\infty \in \mathcal{C}_E([0,1])$. We have that $u_\infty(0) = x_0$. There is $g_n \in \mathcal{S}^2_\Gamma$ satisfying (*) $\dot{u}_n(t) \in -A(t)u_n(t) + g_n(t)$ and (**) $g_n(t) \in G(t, u_n(t))$ for almost every $t \in [0,1]$. By extracting subsequences, we may suppose that (g_n) converges $\sigma(L^2_E, L^2_{E'})$ to $g_\infty \in \mathcal{S}^2_K$ and $\dot{u}_n$ converges $\sigma(L^2_E, L^2_{E'})$ to $\dot{u}_\infty$. By (*) and Lemma 2.3, we have that $\dot{u}_\infty(t) \in A(t)u_\infty(t) + g_\infty(t)$ for almost every $t \in [0,1]$. Using (**) and applying the closure theorem in ([9], Theorem VI-4), gives $g_\infty(t) \in G(t, u_\infty(t))$ for almost every $t \in [0,1]$. This proves that u_∞ is a strong solution of (P_0).

□

3. Applications

We present some typical problems in Optimal Control theory. Let Z be a compact metric space, $k(Z)$ is the set of all compact subsets of Z, and let $\mathcal{M}^1_+(Z)$ be the set of all probability Radon measures on Z. It is well-known that $\mathcal{M}^1_+(Z)$ is a compact metrizable space for the $\sigma(\mathcal{C}(Z)', \mathcal{C}(Z))$-topology. We need first a density result.

Proposition 3.1. *(Castaing-Valadier 1971, unpublished.) Let $(T, \mathcal{T}, \mu)$ be a complete probability space and let Z be a compact metric space. Let $\Gamma : T \to \mathcal{B}(Z)$ be a multifunction with measurable graph and let*

$$\Sigma(t) := \{\nu \in \mathcal{M}^1_+(Z) : \nu(\Gamma(t)) = 1\}$$

for all $t \in T$. Let $\lambda \in \mathcal{S}_\Sigma$ (where $\mathcal{S}_\Sigma$ is the set of all $\mathcal{T}$-measurable selections of Σ) and $\sigma_\lambda = \int_T \delta_t \otimes \lambda_t\, \mu(dt)$. Let $\varepsilon > 0$, $(g_i : i \in I)$ and $(h_i : i \in I)$ two finite sequences in $L^1(T,\mu)$ and $\mathcal{C}(Z)$ respectively. If μ is nonatomic, then there exists a $\mathcal{T}$-measurable selection of Γ, ζ, such that

$$|\langle \sigma_\lambda, g_i \otimes h_i \rangle - \langle \sigma_\zeta, g_i \otimes h_i \rangle| \leq \varepsilon, \ \forall i \in I$$

where $\sigma_\zeta = \int_T \delta_t \otimes \delta_{\zeta(t)} \, \mu(dt)$.

Proof. Let us set $M := \sup_{i \in I} \int_T |g_i(t)| \, \mu(dt)$. There exists a finite Borel partition $(\Omega_j, j \in J)$ of Z such that for any pair z, z' in Ω_j, the following holds $|h_i(z) - h_i(z')| \leq \frac{\varepsilon}{M}$, $\forall i \in I$. Let $\zeta_j : \Omega \to Z$ be a $\mathcal{T}$-mesurable mapping defined by: $\zeta_j(t) \in \Gamma(t) \cap \Omega_j$ if $\Gamma(t) \cap \Omega_j \neq \emptyset$, and $\zeta_j(t) \in \Gamma(t)$ otherwise; whose existence is ensured by Von Neumann-Aumann measurable selection theorem. Let us set $\nu_t := \Sigma_{j \in J} \lambda_t(\Omega_j) \delta_{\zeta_j(t)}$, $\forall t \in T$. Then it is obvious that $t \mapsto \nu_t$ is scalarly $\mathcal{T}$-measurable (alias Young measure). Then $\langle \sigma_\nu, g_i \otimes h_i \rangle = \Sigma_{j \in J} \int_T \lambda_t(\Omega_j) g_i(t) h_i(\zeta_j(t)) \, \mu(dt)$. Further we have $\langle \sigma_\lambda - \sigma_\nu, g_i \otimes h_i \rangle = \int_T g_i(t) \langle \lambda_t - \nu_t, h_i \rangle \, \mu(dt)$ with $\langle \lambda_t - \nu_t, h_i \rangle := \int_{\Gamma(t)} [h_i(z) - \Sigma_{j \in J} 1_{\Omega_j}(z) h_i(u_j(t))] \, \lambda_t(dz)$. As the integrand in the preceding integral is estimated by $\frac{\varepsilon}{M}$, we get $|\langle \sigma_\lambda - \sigma_\nu, g_i \otimes h_i \rangle| \leq \varepsilon$ for all $i \in I$. As μ is nonatomic, by Ljapunov's theorem, there exists a $\mathcal{T}$-measurable partition $(T_j, j \in J)$ of T such that

$$\int_T \lambda_t(\Omega_j) g_i(t) h(\zeta_j(t)) \, \mu(dt) = \int_{T_j} g_i(t) h_i(\zeta_j(t)) \, \mu(dt),$$

for all $i \in I$ and $j \in J$. Let us set $\zeta(t) = \zeta_j(t)$ if $t \in T_j$, $j \in J$ and $\sigma_\zeta = \int_T \delta_t \otimes \delta_{\zeta(t)} \, \mu(dt)$. Then we have

$$\langle \sigma_\nu, g_i \otimes h_i \rangle = \langle \sigma_\zeta, g_i \otimes h_i \rangle,$$

for all $i \in I$. It follows that

$$|\langle \sigma_\lambda, g_i \otimes h_i \rangle - \langle \sigma_\zeta, g_i \otimes h_i \rangle| \leq \varepsilon,$$

for all $i \in I$. □

In view of the preceding result, the set $\mathcal{S}_\Gamma$ of all Lebesgue-measurable selections (alias original controls) of Γ is dense for the $\sigma(L^\infty_{C(Z)'}([0,1]), L^1_{C(Z)}([0,1]))$ topology in the set $\mathcal{S}_\Sigma$ of Lebesgue-measurable selections (alias relaxed controls) of the Lebesgue-measurable multifunction $\Sigma : [0,1] \to \mathcal{M}^1_+(Z)$ defined by

$$\Sigma(t) := \{\nu \in \mathcal{M}^1_+(Z) : \nu(\Gamma(t)) = 1\}$$

for all $t \in [0,1]$.

Let us consider a mapping $g : [0,1] \times E \times Z \to E$ satisfying:

(i) for every fixed $t \in [0,1]$, $g(t,.,.)$ is continuous on $E \times Z$,

(ii) for every $(x,z) \in E \times Z$, $g(.,x,z)$ is Lebesgue-measurable on $[0,1]$,

(iii) for every $\eta > 0$, there exists $l(\eta) > 0$ such that $||g(t,x,z) - g(t,y,z)|| \leq l(\eta)||x - y||$ for all $t \in [0,1]$ and for all $(x,y,z) \in \overline{B}_E(0,\eta) \times \overline{B}_E(0,\eta) \times Z$ where $\overline{B}_E(0,\eta)$ denotes the closed ball in E with center 0 and radius η,

(iv) there exists $c > 0$ such that $g(t,x,Z) \subset c\overline{B}_E$ for all $(t,x) \in [0,1] \times E$.

Let us set

$$h(t,x,\nu) := \int_Z g(t,x,z)\nu(dz),$$

for all $(t,x,\nu) \in [0,1] \times E \times \mathcal{M}_+^1(Z)$. Then h inherits the properties of the function g. Namely, we have

$(i)'$ for every fixed $t \in [0,1]$, $h(t,.,.)$ is continuous on $E \times \mathcal{M}_+^1(Z)$,

$(ii)'$ for every $(x,\nu) \in E \times \mathcal{M}_+^1(Z), h(.,x,\nu)$ is Lebesgue-measurable on $[0,1]$,

$(iii)'$ for every $\eta > 0$, there exists $l(\eta) > 0$ such that $||h(t,x,z) - h(t,y,z)|| \leq l(\eta)||x - y||$ for all $t \in [0,1]$ and for all $(x,y,\nu) \in \overline{B}_E(0,\eta) \times \overline{B}_E(0,\eta) \times \mathcal{M}_+^1(Z)$ where $\overline{B}_E(0,\eta)$ denotes the closed ball in E with center 0 and radius η,

$(iv)'$ there exists $c > 0$ such that $h(t,x,\mathcal{M}_+^1(Z)) \subset c\overline{B}_E$ for all $(t,x) \in [0,1] \times E$.

In the remainder of we will assume that $\Gamma : [0,1] \to Z$ is a *compact valued* measurable multifunction. Now let $A(t)$ be a m-accretive operator in E satisfying $(H_1), (H_2), (H_3)$ with $\overline{D} := \overline{D(A(t))}$ ball-compact. We aim to compare the strong solutions set $(\mathcal{S}_\mathcal{O}) := \{u_\zeta : \zeta \in \mathcal{S}_\Gamma\}$ of the evolution equation:

$$(P_\mathcal{O}) \begin{cases} \dot{u}_\zeta(t) \in -A(t)u_\zeta(t) + g(t, u_\zeta(t), \zeta(t)) \ a.e \ t \in [0,1], \\ u_\zeta(0) = x_0 \in \overline{D}, \end{cases}$$

where ζ belongs to the set $\mathcal{S}_\Gamma$ of all original controls, with the strong solutions set $(\mathcal{S}_\mathcal{R}) := \{u_\lambda : \lambda \in \mathcal{S}_\Sigma\}$ of the evolution equation

$$(P_\mathcal{R}) \begin{cases} \dot{u}_\lambda(t) \in -A(t)u_\lambda(t) + \displaystyle\int_{\Gamma(t)} g(t, u_\lambda(t), z)\lambda_t(dz) \ a.e \ t \in [0,1], \\ u_\lambda(0) = x_0 \in \overline{D}, \end{cases}$$

where λ belongs to the set $\mathcal{S}_\Sigma$ of all relaxed controls. In view of Propositions 2.9-2.10 b), for each $\lambda \in \mathcal{S}_\Sigma$, there is a unique solution u_λ to the problem

$$(P_\mathcal{R}) \begin{cases} \dot{u}_\lambda(t) \in -A(t)u_\lambda(t) + \displaystyle\int_{\Gamma(t)} g(t, u_\lambda(t), z)\lambda_t(dz) \ a.e \ t \in [0,1], \\ u_\lambda(0) = x_0 \in \overline{D}. \end{cases}$$

Let us set $H(t,x) := h(t,x,\Sigma(t)), \forall (t,x) \in [0,1] \times E$. Then it easy to check that H is a separately scalarly measurable and separately scalarly upper semi-continuous convex compact valued multifunction satisfying $H(t,x) \subset c\overline{B}_E$ for all $(t,x) \in [0,1] \times E$. By Proposition 2.10 b) the strong solutions set of

$$\begin{cases} \dot{u}(t) \in -A(t)u(t) + H(t,u(t)) \ a.e \ t \in [0,1], \\ u(0) = x_0 \in \overline{D}, \end{cases}$$

is compact in $\mathcal{C}_E([0,1])$. Using measurable selection theorem, we see that $(\mathcal{S}_{\mathcal{R}})$ coincides with the strong solutions set of the preceding differential inclusion. So we conclude that the solutions set $\{u_\lambda : \lambda \in \mathcal{S}_\Sigma\}$ of $(P_{\mathcal{R}})$ is compact in $\mathcal{C}_E([0,1])$. It is obvious that the solutions set $\{u_\zeta : \zeta \in \mathcal{S}_\Gamma\}$ is a nonempty subset of $\{u_\lambda : \lambda \in \mathcal{S}_\Sigma\}$. Now comes a relaxation property for the problem under consideration.

Theorem 3.2. *Assume that E is a separable Hilbert space, $A(t)$ is a maximal monotone operator in E, and $(H_1), (H_2), (H_3)$ are satisfied with $\overline{D} := \overline{D(A(t))}$ ball-compact. Then the solution sets $(\mathcal{S}_{\mathcal{O}}) = \{u_\zeta : \zeta \in \mathcal{S}_\Gamma\}$ is dense with respect to the uniform norm in the compact solutions set $(\mathcal{S}_{\mathcal{R}}) = \{u_\lambda : \lambda \in \mathcal{S}_\Sigma\}$.*

It is clear that Theorem 3.3 follows from

Lemma 3.3. *The graph of the single-valued mapping $\lambda \mapsto u_\lambda$ defined on the $\sigma(L^\infty_{C(Z)'}([0,1]), L^1_{C(Z)}([0,1]))$ compact set $\mathcal{S}_\Sigma$ is closed.*

Proof. Let $\lambda^n \to \lambda^\infty$ for the $\sigma(L^\infty_{C(Z)'}([0,1]), L^1_{C(Z)}([0,1]))$ topology and $u_{\lambda^n} \to u_\infty$ in $\mathcal{C}_H([0,1])$ where u_{λ^n} $(n \in \mathbb{N} \cup \{\infty\})$ is the unique absolutely continuous solution of the equation

$$\begin{cases} \dot{u}_{\lambda^n}(t) \in -A(t)u_{\lambda^n}(t) + \displaystyle\int_{\Gamma(t)} g(t, u_{\lambda^n}(t), z)\lambda^n_t(dz) \ a.e \ t \in [0,1], \\ u_{\lambda^n}(0) = x_0 \in \overline{D}, \end{cases}$$

then $u_\infty = u_{\lambda^\infty}$. For simplicity we set

$$h_\lambda(t,x) = \int_{\Gamma(t)} g(t,x,z)\lambda_t(dz) = \int_Z g(t,x,z)\lambda_t(dz) = h(t,x,\lambda_t)$$

for all $(t,x,\lambda) \in [0,1] \times E \times \mathcal{S}_\Sigma$. We have

$$h_{\lambda^n}(t, u_{\lambda^n}(t)) - \dot{u}_{\lambda^n}(t) \in A(t)u_{\lambda^n}(t),$$

and

$$h_{\lambda^\infty}(t, u_{\lambda^\infty}(t)) - \dot{u}_{\lambda^\infty}(t) \in A(t)u_{\lambda^\infty}(t),$$

for a.e $t \in [0, 1]$. Since $A(t)$ is monotone,

$$\langle u_{\lambda^n}(t) - u_{\lambda^\infty}(t), h_{\lambda^n}(t, u_{\lambda^n}(t)) - \dot{u}_{\lambda^n}(t) - h_{\lambda^\infty}(t, u_{\lambda^\infty}(t)) + \dot{u}_{\lambda^\infty}(t) \geq 0,$$

a.e in $[0, 1]$. So

$$\frac{1}{2}\frac{d}{dt}||u_{\lambda^n}(t) - u_{\lambda^\infty}(t)||^2 \leq \langle u_{\lambda^n}(t) - u_{\lambda^\infty}(t), h_{\lambda^n}(t, u_{\lambda^n}(t)) - h_{\lambda^\infty}(t, u_{\lambda^\infty}(t))\rangle,$$

a.e in $[0, 1]$. Integrating on $[0, t]$ gives

$$\frac{1}{2}||u_{\lambda^n}(t) - u_{\lambda^\infty}(t)||^2$$
$$\leq \int_0^t \langle u_{\lambda^n}(s) - u_{\lambda^\infty}(s), h_{\lambda^n}(s, u_{\lambda^n}(s)) - h_{\lambda^\infty}(s, u_{\lambda^\infty}(s))\rangle\, ds.$$

Let us set

$$L_n(t) = \int_0^t \langle u_{\lambda^n}(s) - u_{\lambda^\infty}(s), h_{\lambda^n}(s, u_{\lambda^n}(s)) - h_{\lambda^\infty}(s, u_{\lambda^\infty}(s))\rangle\, ds.$$

Then $L_n(t) = L_n^1(t) + L_n^2(t) + L_n^3(t)$ where

$$L_n^1(t) = \int_0^t \langle u_{\lambda^n}(s) - u_{\lambda^\infty}(s), h_{\lambda^n}(s, u_{\lambda^n}(s)) - h_{\lambda^n}(s, u_{\lambda^\infty}(s))\rangle\, ds,$$

$$L_n^2(t) = \int_0^t \langle u_{\lambda^n}(s) - u_\infty(t), h_{\lambda^n}(s, u_{\lambda^\infty}(s)) - h_{\lambda^\infty}(s, u_{\lambda^\infty}(s))\rangle\, ds,$$

$$L_n^3(t) = \int_0^t \langle u_\infty(s) - u_{\lambda^\infty}(s), h_{\lambda^n}(s, u_{\lambda^\infty}(s)) - h_{\lambda^\infty}(s, u_{\lambda^\infty}(s))\rangle\, ds.$$

As $||h_\lambda(t, x)|| \leq c$ for all $\lambda \in S_\Sigma$ and for all $(t, x) \in [0, 1] \times H$, using $(iv)'$, we get the estimation

$$|L_n^2(t)| \leq 2c||u_{\lambda^n} - u_\infty||_{C_H([0,1])}.$$

Thus $L_n^2(t) \to 0$ uniformly in $[0, 1]$, when $n \to \infty$. Similarly by (iv) the integrand

$$f(s, z) := \langle u_\infty(s) - u_{\lambda^\infty}(s), g(s, u_{\lambda^\infty}(s), z)\rangle$$

is estimated by

$$|f(s, z)| \leq c||u_\infty - u_{\lambda^\infty}||_{C_H([0,1])},$$

for all $(s, z) \in [0, 1] \times Z$. Hence $f \in L^1_{C(Z)}([0, 1])$. As (λ^n) converges $\sigma(L^\infty_{C(Z)'}, L^1_{C(Z)})$ to λ^∞, it is immediate that for every $t \in [0, 1]$,

$$\int_0^t [\int_Z f(s,z)\lambda_s^n(dz)]\,ds \to \int_0^t [\int_Z f(s,z)\lambda_s^\infty(dz)]\,ds$$

when $n \to \infty$. So $\lim_{n\to\infty} L_n^3(t) = 0$. Further there is $\eta > 0$ such that

$$\sup\{||u_\lambda||_{C_H([0,1])} : \lambda \in S_\Sigma\} < \eta < \infty.$$

Using $(iv)'$, there is $l(\eta) > 0$ such that

$$|L_n^1(t)| \leq \int_0^t l(\eta)||u_{\lambda^n}(s) - u_{\lambda^\infty}(s)||^2\,ds.$$

Finally we get

$$\frac{1}{2}||u_{\lambda^n}(t) - u_{\lambda^\infty}(t)||^2 \leq L_n^2(t) + L_n^3(t) + \int_0^t l(\eta)||u_{\lambda^n}(s) - u_{\lambda^\infty}(s)||^2\,ds.$$

As the functions $L_n^2(.)$ and $L_n^3(.)$ are continuous with $L_n^2(t) \to 0$ and $L_n^3(t) \to 0$, by Gronwall's lemma we conclude that

$$u_{\lambda^n}(t) \to u_{\lambda^\infty}(t)$$

for all $t \in [0,1]$. So, $u_{\lambda^\infty} = u_\infty$ and the set $\{u_\lambda : \lambda \in \mathcal{S}_\Sigma\}$ is compact in $C_H([0,1])$. $\square$

There is variant of Theorem 3.2.

Example 3.4. Let E be a separable Hilbert space, V be a convex weakly compact subset of H and let $\Delta : [0,1] \to V$ be a convex weakly compact valued upper semicontinuous multifunction. In view of ([9], Theorem IV.16) the graph of multifunction $ext(\Delta) : t \to ext(\Delta(t))$, where $ext(\Delta(t))$ denotes the set of all extreme points of $\Delta(t)$, is a Borel subset in $[0,1] \times V$. So, one can apply Proposition 3.1 by taking $(T, \mathcal{T}, \mu) = ([0,1], \mathcal{L}([0,1]); \lambda)$, $Z = V_{weak}$, $\Gamma = ext(\Delta)$ and $\Sigma = \Psi$ where

$$\Psi(t) := \{\nu \in \mathcal{M}_+^1(V) : \nu(ext(\Delta(t)) = 1\},$$

for all $t \in [0,1]$. Repeating the above arguments, we conclude that the strong solutions set $(\mathcal{S}_{ext(\Delta)})$ of the equation

$$(P_{ext(\Delta)}) \begin{cases} \dot{u}_\zeta(t) \in -A(t)u_\zeta(t) + g(t, u_\zeta(t), \zeta(t)) \ a.e\ t \in [0,1], \\ u_\zeta(0) = x_0 \in \overline{D}, \end{cases}$$

where ζ belongs to the set of all Lebesgue-measurable selections of the multifunction $ext(\Delta(.))$ is dense for the uniform convergence in the strong solutions set $(\mathcal{S}_\Psi)$ of

$$(P_\Psi)\begin{cases}\dot{u}_\lambda(t) \in -A(t)u_\lambda(t) + \displaystyle\int_{ext(\Delta(t))} g(t, u_\lambda(t), u)\lambda_t(du) \ a.e\ t \in [0,1],\\ u_\lambda(0) = x_0 \in \overline{D},\end{cases}$$

where λ is a Lebesgue-measurable selection of the multifunction Ψ defined above.

In the same vein, we have the following which concerns an analogous relaxation property for a quasi-autonomuous evolution equation governed by an m-acretive operator.

Example 3.5. Let $A(t)$ be an $cc(E)\cup\{\emptyset\}$ m-accretive operator satisfying (H_1), $(H_2), (H_3)$ with $\overline{D} := \overline{D(A(t))}$ ball-compact, K be a convex weakly compact subset of E and let $F : [0,1] \to K$ be a convex weakly compact valued upper semicontinuous multifunction. As above the graph of multifunction $ext(F)$: $t \to ext(F(t))$, where $ext(F(t))$ denotes the set of all extreme points of $F(t)$, is a Borel subset in $[0,1]\times K$. We claim that the strong solutions set $(\mathcal{S}_{ext(F)})$ of the equation

$$(P_{ext(F)})\begin{cases}\dot{u}_f(t) \in -A(t)u_f(t) + f(t) \ a.e\ t \in [0,1],\\ u_f(0) = x_0 \in \overline{D},\end{cases}$$

where f belongs to the set $S^\infty_{ext(F)}$ of all bounded Lebesgue-measurable selections of the multifunction $ext(F)$ is dense for the topology of uniform convergence in the the compact solutions set $(\mathcal{S}_F)$ of the equation

$$(P_F)\begin{cases}\dot{u}_f(t) \in -A(t)u_f(t) + f(t) \ a.e\ t \in [0,1],\\ u_f(0) = x_0 \in \overline{D},\end{cases}$$

where f belongs to the set S^∞_F all bounded Lebesgue-measurable selections of the multifunction F. First, observe that, in view of Corollary 2.6 and Proposition 2.9, for each $f \in S^\infty_F$, there exists a unique strong solution u_f for the equation

$$\begin{cases}\dot{u}_f(t) \in -A(t)u_f(t) + f(t) \ a.e\ t \in [0,1],\\ u_f(0) = x_0 \in \overline{D}.\end{cases}$$

Second, it is well-known (see e.g [9]) that $S^\infty_{ext(F)}$ is dense in S^∞_F for the $\sigma(L^\infty_E, L^1_{E'})$-topology, using Ljapunov's theorem. By Theorem 2.10 a), $(\mathcal{S}_F)$ is relatively compact in $\mathcal{C}_E([0,1])$. So, it is enough to remark that the graph of the mapping $f \mapsto u_f$ from S^∞_F equipped with the topology $\sigma(L^\infty_E, L^1_{E'})$-topology into $\mathcal{C}_E([0,1])$ equipped with the topology of uniform convergence is closed, repeating the arguments given in the proof of Proposition 2.10 b). Details are left to the reader. $\square$

Comments. If E is a finite dimensional space, ball-compactness assumption on $\overline{D}$ in Theorem 2.4 is superfluous. The study of undelayed differential inclusions governed by m-accretive operators arises from Economics [10, 16], Mechanics [20, 21] and Control Theory [8, 14, 17, 29, 30]. Other results in relaxed control theory for both second order ordinary differential equations and second order multivalued differential equations appear in a forthcoming paper by Azzam-Castaing-Thibault.

To end this paper, we present now the existence of absolutely continuous solutions for a special class of evolution equation in a separable Hilbert space H. Here compactness assumption is unnecessary by constrast to the results presented above.

Proposition 3.6. *Let H be a separable Hilbert space. Let $f : [0,1] \times H \to \mathbb{R}$ satisfying:*

(1) for every $x \in H$, $f(.,x)$ is Lebesgue-measurable,

(2) there exists $\beta \in L^2_{\mathbb{R}^+}([0,1])$ such that for all $t \in [0,1]$ and for all x, y in H

$$|f(t,x) - f(t,y)| \leq \beta(t)||x-y||,$$

(3) for every $t \in [0,1]$, the function $f(t,.)$ is convex.

Then, for any $a \in H$, there is a unique absolutely continuous solution to the equation

$$\begin{cases} \dot{u}(t) \in -\partial f_t(u(t)) \ a.e\ t \in [0,1], \\ u(0) = a \in H. \end{cases}$$

Proof. Let $A(t) = \partial f_t$. According to (1), (2), (3) and the example just after Lemma 2.3, the maximal monotone operator $A(t)$ satisfies assumption (H_3). In particular, the mappings $t \mapsto J_\lambda A(t)x$ and $t \mapsto A_\lambda(t)x$ are measurable on $[0,1]$ for every $\lambda > 0$ and every $x \in H$. Note that for each $\lambda > 0$ the equation

$$\begin{cases} \dfrac{du_\lambda}{dt}(t) + A_\lambda(t)u_\lambda(t) = 0, \\ u_\lambda(0) = a \in H, \end{cases}$$

admits a unique absolutely continuous solution because $A_\lambda(t)x$ is separately measurable and separately Lipschitzean and satisfies the estimate $||A_\lambda(t)x|| \leq \beta(t)$, for all $\lambda > 0$ and for all $(t,x) \in [0,1] \times H$.

Now let (λ_n) be a decreasing sequence of positive real numbers such that $\lambda_n \downarrow 0$. For each integer n there exists a absolutely continuous solution $u_{\lambda_n} : [0,1] \to H$ to the equation:

$$\begin{cases} \frac{du_{\lambda_n}}{dt}(t) + A_{\lambda_n}(t)u_{\lambda_n}(t) = 0, \\ u_{\lambda_n}(0) = a. \end{cases}$$

Furthermore we have the estimate

$$\forall n \in \mathbb{N}, \ ||\dot{u}_{\lambda_n}||_{L^2_H([0,1])} \leq ||\beta||_{L^2_{\mathbb{R}}([0,1])}.$$

Let m and n be any positive integers. We have

$$\begin{aligned} &\frac{d}{dt}||u_{\lambda_n}(t) - u_{\lambda_m}(t)||^2 \\ &= 2\langle u_{\lambda_n}(t) - u_{\lambda_m}(t), \dot{u}_{\lambda_n}(t) - \dot{u}_{\lambda_m}(t)\rangle \\ &= 2\langle \lambda_n A_{\lambda_n}(t)u_{\lambda_n}(t) - \lambda_m A_{\lambda_m}(t)u_{\lambda_m}(t), \dot{u}_{\lambda_n}(t) - \dot{u}_{\lambda_m}(t)\rangle \\ &\quad + 2\langle J_{\lambda_n}(t)u_{\lambda_n}(t) - J_{\lambda_m}(t)u_{\lambda_m}(t), \dot{u}_{\lambda_n}(t) - \dot{u}_{\lambda_m}(t)\rangle. \end{aligned} \tag{3.6.1}$$

Here $J_\lambda(t)$ stands for $J_\lambda A(t)$. Notice that

$$-\dot{u}_{\lambda_m}(t) = A_{\lambda_m}(t)u_{\lambda_m}(t) \in A(t)J_{\lambda_m}(t)u_{\lambda_m}(t) \tag{3.6.2}$$

and

$$-\dot{u}_{\lambda_n}(t) = A_{\lambda_n}(t)u_{\lambda_n}(t) \in A(t)J_{\lambda_n}(t)u_{\lambda_n}(t). \tag{3.6.3}$$

From (3.6.2), (3.6.3) and the monotonicity of $A(t)$, it follows that

$$\langle J_{\lambda_n}(t)u_{\lambda_n}(t) - J_{\lambda_m}(t)u_{\lambda_m}(t), \dot{u}_{\lambda_n}(t) - \dot{u}_{\lambda_m}(t)\rangle \leq 0. \tag{3.6.4}$$

Using (3.6.4) and coming back to (3.6.1) we get

$$\begin{aligned} &\frac{d}{dt}||u_{\lambda_n}(t) - u_{\lambda_m}(t)||^2 \\ &\leq 2\langle \lambda_n A_{\lambda_n}(t)u_{\lambda_n}(t) - \lambda_m A_{\lambda_m}(t)u_{\lambda_m}(t), \dot{u}_{\lambda_n}(t) - \dot{u}_{\lambda_m}(t)\rangle \\ &= -2\langle \lambda_n \dot{u}_{\lambda_n}(t) - \lambda_m \dot{u}_{\lambda_m}(t), \dot{u}_{\lambda_n}(t) - \dot{u}_{\lambda_m}(t)\rangle. \end{aligned} \tag{3.6.5}$$

Integrating inequality (3.6.5) on $[0, t]$ gives

$$\begin{aligned} &||u_{\lambda_n}(t) - u_{\lambda_m}(t)||^2 \\ &\leq -2\langle \lambda_n \dot{u}_{\lambda_n}(.) - \lambda_m \dot{u}_{\lambda_m}(.), \dot{u}_{\lambda_n}(.) - \dot{u}_{\lambda_m}(.)\rangle_{\langle L^2_H([0,t]), L^2_H([0,t])\rangle}. \end{aligned} \tag{3.6.6}$$

Consequently we get

$$\begin{aligned} &\forall\, m, n, \\ &\langle \lambda_n \dot{u}_{\lambda_n}(.) - \lambda_m \dot{u}_{\lambda_m}(.), \dot{u}_{\lambda_n}(.) - \dot{u}_{\lambda_m}(.)\rangle_{\langle L^2_H([0,t]), L^2_H([0,t])\rangle} \leq 0. \end{aligned} \tag{3.6.7}$$

Since the sequence $(\dot{u}_{\lambda_n}(.))$ is bounded in $L^2_H([0,1])$, by (3.6.7) and [11], $(\dot{u}_{\lambda_n}(.))$ converges in the Hilbert space $L^2_H([0,1])$. Let v be the limit of $(\dot{u}_{\lambda_n}(.))$ and let, for all $t \in [0,1]$

$$u(t) = a + \int_0^t v(s)\,ds.$$

Then $u = \lim_{n\to\infty} u_{\lambda_n}(.)$ in $\mathcal{C}_H([0,1])$ because the mapping $f \mapsto \int_0^1 f(s)\,ds$ from $L^2_H([0,1])$ to $\mathcal{C}_H([0,1])$ is continuous for the norm topologies. Now we finish the proof by showing that u satisfies

$$\begin{cases} \dot{u}(t) \in -A(t)u(t) \ a.e\ t \in [0,1], \\ u(0) = a \in H. \end{cases}$$

By (3.6.3) we have

$$-\dot{u}_{\lambda_n}(t) = A_{\lambda_n}(t)u_{\lambda_n}(t) \in A(t)J_{\lambda_n}(t)u_{\lambda_n}(t). \tag{3.6.8}$$

But

$$\begin{aligned} &\|J_{\lambda_n}(t)u_{\lambda_n}(t) - u(t)\| \\ &\le \|J_{\lambda_n}(t)u_{\lambda_n}(t) - u_{\lambda_n}(t)\| + \|u_{\lambda_n}(t) - u(t)\|, \end{aligned} \tag{3.6.9}$$

with

$$\begin{aligned} \|J_{\lambda_n}(t)u_{\lambda_n}(t) - u_{\lambda_n}(t)\| &= \lambda_n\|A_{\lambda_n}(t)u_{\lambda_n}(t)\| \\ &= \lambda_n\|\dot{u}_{\lambda_n}(t)\| \\ &\le \lambda_n\beta(t). \end{aligned} \tag{3.6.10}$$

As $\lambda_n\beta(t) \to 0$, from (3.6.9) and (3.6.10) we see that

$$\|J_{\lambda_n}(t)u_{\lambda_n}(t) - u(t)\| \to 0. \tag{3.6.11}$$

By extracting a subsequence, we may suppose that $(\dot{u}_{\lambda_n}(.))$ converges a.e. to v, for H endowed with the norm topology. As the graph of $A(t) = \partial f_t$ is sequentially strongly×weakly closed, from (3.6.8) and (3.6.11), we finally get $-v(t) \in A(t)u(t)$ a.e. Since $\dot{u} = v$ a.e, the proof is therefore complete because the uniqueness is straightforward, using the monotonicity of $A(t)$ (see also Proposition 2.9). □

Now we give an application of Proposition 3.6 which allows us to recover the existence of absolutely continuous solutions for the convex sweeping process in Hilbert spaces [21].

Proposition 3.7. *Let $C : [0,1] \to H$ be a closed convex valued multifunction satisfying: there is a positive number $k > 0$ such that, for all $x, y \in H$ and for all $s, t \in [0,1]$*

$$|d(x, C(t)) - d(y, C(s))| \leq ||x - y|| + k|t - s|.$$

Let $a \in C(0)$. Then there is a unique absolutely continuous solution $u : [0,1] \to H$ to

$$\begin{cases} \dot{u}(t) \in -\partial[kd_{C(t)}](u(t)) \ a.e \ t \in [0,1], \\ u(0) = a, \ \ u(t) \in C(t), \ \forall t \in [0,1]. \end{cases}$$

Proof. By Proposition 3.6, there is a unique absolutely continuous solution $u : [0,1] \to H$ to

$$\begin{cases} \dot{u}(t) \in -\partial[kd_{C(t)}](u(t)) \ a.e \ t \in [0,1], \\ u(0) = a \in C(0). \end{cases}$$

To finish the proof it is enough to apply Theorem 1.3 in [26] asserting that any absolutely continuous solution $u(.)$ of the preceding equation satisfies the inclusion $u(t) \in C(t)$ for all $t \in [0,1]$. □

Corollary 3.8. *Let $C : [0,1] \to H$ be a closed convex valued multifunction satisfying: there is a positive number $k > 0$ such that, for all $x, y \in H$ and for all $s, t \in [0,1]$*

$$|d(x, C(t)) - d(y, C(s))| \leq ||x - y|| + k|t - s|.$$

Let $a \in C(0)$. Then there is a unique absolutely continuous solution $z : [0,1] \to H$ to

$$\begin{cases} \dot{z}(t) \in -N_{C(t)}(z(t)) \ a.e \ t \in [0,1], \\ z(0) = a, \ \ z(t) \in C(t), \ \forall t \in [0,1]. \end{cases}$$

Proof. Recall that $N_{C(t)}(z(t))$ denotes the normal cone of $C(t)$ at the point $z(t)$. Let z(.) be the unique absolutely continuous solution $z : [0,1] \to H$ to

$$\begin{cases} \dot{z}(t) \in -\partial[kd_{C(t)}](z(t)) \ a.e \ t \in [0,1], \\ z(0) = a \in C(0), \ \ z(t) \in C(t), \ \forall t \in [0,1], \end{cases}$$

whose existence is ensured by Proposition 3.7. Then, by ([25], Proposition II.12), we get

$$\partial[kd_{C(t)}](z(t)) \subset N_{C(t)}(z(t)),$$

and hence $-\dot{z}(t) \in N_{C(t)}(z(t))$. □

References

[1] Barbu, V.: Nonlinear semigroups and differential equations in Banach spaces. Noordhoff Int. Publ., Leyden (1976)

[2] Benabdellah, H., Faik, A.: Perturbations convexes et nonconvexes des équations d'évolution. Portugaliae Matematica **53(2)**, 187-208 (1996)

[3] Brezis, H.: Opérateurs maximaux monotones et semi-groupes de contractions dans un espace de Hilbert. North Holland 1979

[4] Castaing, C., Faik, A., Salvadori, A.: Evolution equations governed by m-accretive and subdifferential operators. International Journal of Applied Mathematics **2(9)**, 1005-1026 (2000)

[5] Castaing, C., Ibrahim, A.G.: Functional differential inclusion on closed sets in Banach spaces. Adv. Math. Econ **2**, 21-39 (2000)

[6] Castaing, C., Monteiro Marques, M.D.P.: A multivalued version of Scorza-Dragoni's theorem with an application to normal integrands. Bull. Pol. Acad. Sci. Mathematics **42**, 133-140 (1994)

[7] Castaing, C., Monteiro Marques, M.D.P.: Topological properties of solutions sets for sweeping processes with delay. Portugaliae Mathematica **54(4)**, 485-507 (1997)

[8] Castaing, C., Salvadori, A., Thibault, L.: Functional evolution equations governed by nonconvex sweeeping process. Journal of Nonlinear and Convex Analysis **2(2)**, 217-241 (2001)

[9] Castaing, C., Valadier, M.: Convex analysis and measurable multifunctions. Lecture Notes in Math. **580**, Springer-Verlag, Berlin (1977)

[10] Cornet, B.: Existence of slow solutions for a class of differential inclusions. JMMA **96**, 130-147 (1983)

[11] Crandall, M.G., Pazy, M.: Semi-groups of nonlinear contractions and dissipative sets on general Banach spaces. J. Funct. Anal. **3**, 376-418 (1969)

[12] Evans, L.C.: Nonlinear evolution equation in an arbitrary Banach space. Israel J. Math. **26(1)**, 1-24 (1977)

[13] Gavioli, A., Malaguti, L.: Viable solutions of differential inclusions with memory in Banach spaces. Portugaliae Mathematica **57(2)**, 203-217 (2000)

[14] Ghouila-Houri, A.: Sur la généralisation de la notion de commande d'un système guidable. Revue d'informatique et de Recherche opérationnelle **4**, 7-32 (1967)

[15] Duc Ha, T.X., Monteiro Marques, M.D.P.: Nonconvex second order differential inclusions with memory. Set-Valued Analysis **3(1)**, 71-86 (1995)

[16] Henry, C.: An existence theorem for a class of differential equations with multivalued right-hand side. J. Math. Anal. Appl. **41**, 179-186 (1973)

[17] Jawhar, A.: Existence de solutions optimales pour les problèmes de contrôle de sytèmes gouvernés par les équations différentielles multivoques. Séminaire d'Analyse convexe, Exposé No 1 1985

[18] Kartsatos, A.G., Parrot, M.E,: A method of lines for a nonlinear abstract functional evolution equation. Trans. Amer. Math. Soc. **286**, 73-89 (1984)

[19] Kunze, M., Monteiro Marques, M.D.P.: B.V solutions to evolution problems with time-dependent domains. Set-Valued Analysis **5**, 57-72 (1997)

[20] Monteiro Marques, M.D.P.: Differential inclusions non nonsmooth mechanical problems, shocks and dry friction. Progress in Nonlinear Differential Equations and Their Applications, Birkhauser **9** (1993)

[21] Moreau, J.J.: Evolution problem asssociated with a moving convex set in a Hilbert space. J. Diff. Eqs. **26**, 347-374 (1977)

[22] Péralba, J.C.: Un problème d'evolution relatif à un opérateur sous-différentiel dépendant du temps. C.R. Acad. Sci. Paris **275** (1972)
[23] Ruess, W.M., Summers, W.H.: Operators semigroups for functional differential equations with delay. Trans. Amer. Math. Soc. **341(2)**, 695-719 (1994)
[24] Tanaka, N.: On the existence of solutions of functional evolution equations. Nonlinear Analysis Methods and Applications **12(10)**, 1087-1104 (1988)
[25] Thibault, L.: Propriétés des sous-différentiels de fonctions localement Lipschitziennes définies sur un espace de Banach séparable. Applications. Thèse, Université Montpellier 1976
[26] Thibault, L.: Sweeping process with regular and nonregular sets. Preprint Montpellier 1999
[27] Valadier, M.: Quelques résultats de base concernant le processus de la rafle. Sém. Anal. Convexe, Montpellier **3** (1988)
[28] Valadier, M.: Young measures. In: Methods of Convex Analysis. Lectures Notes in Math. **1445**, pp.152-188 Springer-Verlag 1990
[29] Warga, J.: Functions of relaxed controls. Siam Journal of Control **5(4)**, 628-641 (1967)
[30] Warga, J.: Optimal control of differential and functional equations. Academic Press, New york 1972
[31] Vrabie, I.L.: Compactness methods for nonlinear evolutions equations. Pitman Monographs and Surveys in Pure and Applied mathematics, Longman Scientific and Technical, John Wiley and Sons, Inc., New York **32** (1987)
[32] Zalinescu, C.: On uniformly convex functions. J. Math. Anal. Appl. **95**, 344-374 (1983)

Adv. Math. Econ. 5, 55–63 (2003)

Advances in
MATHEMATICAL
ECONOMICS

Nonlinear generalizations of theorems on inverse–positive matrices*

Takao Fujimoto[1], **José A. Silva**[2], **and Antonio Villar**[3]

[1] Department of Economic Analysis, University of Alicante, 03071 Alicante, Spain
Department of Economics, Kagawa University, Takamatsu, Kagawa, 760, Japan
(e-mail: takao@ec.kagawa-u.ac.jp)
[2] Department of Economic Analysis, University of Alicante, 03071 Alicante, Spain
[3] Department of Economic Analysis, University of Alicante and IVIE, 03071 Alicante, Spain

Received: August 2, 2002
Revised: September 2, 2002

JEL classification: C67

Mathematics Subject Classification (2000): 15A48

Abstract. Nonlinear extensions of some theorems concerning inverse-positive matrices are made. The method of proof is based on an intuitively natural geometrical consideration.

1. Introduction

The properties of M-matrices are explained in a detailed way by Berman and Plemmons[1], among which the inverse positiveness is well known and important in various fields of sciences. In economics, the positive invertibility of a real square matrix is crucial in discussing the non-substitution theorem for models with joint production, which has been made clear in Herrero and Villar[6]. (See also Schefold[8].) In Fujimoto et al.[5], a necessary and sufficient condition is presented for a real square matrix to have the nonnegative inverse

* Thanks are due to a grant from Ministerio de Ciencia y Tecnologia, under Project BEC2001-0535. This paper was written while Takao Fujimoto was at Department of Fundamental Economic Analysis, University of Alicante, under a sabbatical year professorship given by the Spanish Ministry of Education. He is grateful to hospitality and warm environment there. The authors owe much to careful reading and the comments by the referee.

by generalizing a proposition in Bidard[2, Chap.10, p.113], and a simple proof of the non-substitution theorem is given, using the duality approach by Chander[3]. In this note, we present nonlinear extensions of some theorems concerning inverse-positive matrices. The method of proof is similar to that in [5], and is simple, elementary, and is based on an intuitive geometrical reasoning.

In section 2, we present our assumptions and the propositions. Then in section 3 another type of nonlinear generalization concerning positive invertibility is given, whose linear version has not been stated explicitly so far in the literature. Section 4 explains how this generalization covers the existing results as special cases, raising some numerical examples. Finally in section 5, we conclude with some remarks.

2. Assumptions and theorems

Let us start by explaining notation. The two spaces X and Y are topological vector spaces with their nonempty pointed closed convex cones C and D given respectively. The partial order introduced to X by the cone C is denoted as $x \geq y$ or $y \leq x$ when $x - y \in C$ for $x, y \in C$. The same inequality sign is used for the partial order of Y introduced by D. These cones are assumed to have their nonempty interior int$C$ and int$D$, and we write $x \gg y$ or $y \ll x$ when $x - y \in$int$C$ or int$D$ for $x, y \in C$ or D. One more inequality sign $x > y$ is used when $x - y \in C - \{0\}$ or $x - y \in D - \{0\}$. The boundary of a set S is denoted as bd$S$. The symbol R^n means the real Euclidean space of dimension $n (n \geq 2)$, and R^n_+ the nonnegative orthant of R^n. A given transformation T maps X to Y, and for this T we make the following assumptions.

Assumption 2.1. There exists an $x^* \gg 0$ in X such that $T(x^*) \gg 0$.

Assumption 2.2. If $T(x) \gg 0$ for $x \in C$, then $x \gg 0$.

Assumption 2.3. If $T(x^0) \geq 0$ and $T(x^1) \gg 0$ for x^0, $x^1 \in X$, then $T(kx^0 + (1-k)x^1) \gg 0$ for any scalar k such that $0 < k < 1$.

It is important to note that the assumption 2.2 is made less restrictive by requiring x to be in C. When we consider a simple Leontief model, $T(x) \equiv x - Ax$, which maps R^n into itself with the material input coefficient matrix A being an $n \times n$ nonnegative matrix, the assumption 2.1 requires that the model be *productive*, while the assumption 2.2 is satisfied naturally by the absence of joint production. If we allow for joint production and $T(x) \equiv Bx - Ax$, where B is the $n \times n$ material output coefficient matrix, the assumption 2.2 tells us that every process has to be operated if every commodity is produced in a positive amount, and is called *essentiality* condition in Fujimoto et al.[5]. The assumption 2.3 is satisfied when T is linear. More generally when either (i) the space

Y is R^n, and each component function T_j of T is 'strongly' quasi-concave, i.e., if $T_j(u) > T_j(v)$ then $T_j(ku + (1-k)v) > T_j(v)$ for any scalar k such that $0 < k < 1$, or (ii) T is super-additive, i.e., $T(u+v) \geq T(u) + T(v)$, and positively homogeneous, i.e., $T(kx) = f(k)T(x)$ for a positive scalar k with f being a real non-decreasing function such that $f(k) > 0$ for $k > 0$, the assumption 2.3 is satisfied.

Now we are ready to prove

Proposition 2.1. *Given the assumptions 2.1, 2.2 and 2.3, if* $T(x) \geq 0$ *for* $x \in X$, *then* $x \geq 0$.

Proof. Suppose to the contrary that $x \notin C$. Since the cone C is closed, there exists a scalar k such that $0 < k < 1$ and $\overline{x} \equiv kx + (1-k)x^* \in \mathrm{bd}C$, where x^* is the element given in the assumption 2.1. By the assumption 2.3, $T(\overline{x}) \gg 0$, yielding a contradiction to the assumption 2.2. □

When the assumption 2.2 is strengthened to the following:

Assumption 2.2*. If $T(x) > 0$ for $x \in C$, then $x \gg 0$.

we can prove

Proposition 2.2. *Given the assumptions 2.1, 2.2* and 2.3, if* $T(x) > 0$ *for* $x \in X$, *then* $x \gg 0$.

Proof. By Proposition 2.1, we know that if $T(x) > 0$, then $x \in C$. Thus, the assumption 2.2* guarantees that $x \gg 0$. □

Now we consider one more assumption

Assumption 2.4. If $T(x) = 0$ for $x \in C$, then there exists a $y \notin C$ such that $T(y) = 0$.

This assumption 2.4 is satisfied, for example, when T is homogeneous, i.e., $T(kx) = f(k)T(x)$ for a scalar k with f being a real increasing function on R such that $f(0) = 0$. We now prove

Proposition 2.3. *Let* $T(0) = 0$ *and the assumptions 2.1-2.4 be all satisfied, then the kernel of* T, *KerT* $\equiv \{x | x \in X, T(x) = 0\}$ *is a singleton, consisting of the origin only.*

Proof. Suppose that there exists an $x \in \mathrm{Ker}T$ such that $x \neq 0$. If $x \notin C$, the proof of Proposition 2.1 can be applied to show a contradiction to the assumption 2.2. On the other hand, if $x \in C$, then by the assumption 2.4 we know the existence of y such that $y \notin C$ and $T(y) = 0$, which again leads to a contradiction to the assumption 2.2. □

When T is linear with $X = R^n$ and $Y = R^n$, Propositions 2.1 and 2.3 together are equivalent to the existence of nonnegative inverse of a transformation matrix T.

3. Positive invertibility in a shadow land

In this section, we discuss a nonlinear generalization of inverse-positive matrices which appear in a sort of shadow land of Leontief models. It seems better to start with a numerical example. Let us take up a simple Leontief model in which there are three commodities and three processes to produce those commodities with the material input coefficient matrix A represented as

$$A \equiv \begin{pmatrix} 0.1 & 0.4 & 0.3 \\ 0.2 & 0.1 & 0.2 \\ 0.3 & 0.4 & 0.1 \end{pmatrix}.$$

We consider a mapping $T(x;\lambda) \equiv \lambda x - Ax$, which is thus written as

$$T(x;\lambda) \equiv \begin{pmatrix} \lambda - 0.1 & -0.4 & -0.3 \\ -0.2 & \lambda - 0.1 & -0.2 \\ -0.3 & -0.4 & \lambda - 0.1 \end{pmatrix}.$$

The reciprocal of value λ can be interpreted as the growth factor, i.e., $1+$ rate of steady balanced growth. The Frobenius root λ^* of A is approximately 0.6772, and the eigenvector associated with this eigenvalue is $(0.7069\ldots, 0.4899\ldots, 0.7069\ldots)'$, with the prime $'$ indicating the transposition. It is well known that when $\lambda > \lambda^*$, $T(x;\lambda)$ has its inverse with all the elements positive because A is indecomposable or irreducible. (See Berman and Plemmons[1].)

What happens if λ gets less than λ^* ? At $\lambda = 0.677$, the inverse of $T(x;\lambda)$ is roughly

$$\begin{pmatrix} -1687 & -2339 & -1688 \\ -1170 & -1620 & -1170 \\ -1688 & -2339 & -1687 \end{pmatrix}$$

The inverse of $T(x;\lambda)$ is more generally represented as

$$\begin{pmatrix} \lambda^2 - 0.2\lambda - 0.07 & 0.4\lambda + 0.08 & 0.3\lambda + 0.05 \\ 0.2\lambda + 0.04 & \lambda^2 - 0.2\lambda - 0.08 & 0.2\lambda + 0.04 \\ 0.3\lambda + 0.05 & 0.4\lambda + 0.08 & \lambda^2 - 0.2\lambda - 0.07 \end{pmatrix} / (\lambda^3 - 0.3\lambda^2 - 0.22\lambda - 0.024)$$

Thus, when $0.4 \le \lambda < \lambda^* = 0.6772...$, the negative of $T(x;\lambda)$, i.e., $-T(x;\lambda)$ has its inverse with all the elements nonnegative. For example, at $\lambda = 0.4$, $-T(x;\lambda) = Ax - \lambda x$ is

$$\begin{pmatrix} -0.3 & 0.4 & 0.3 \\ 0.2 & -0.3 & 0.2 \\ 0.3 & 0.4 & -0.3 \end{pmatrix},$$

and its inverse becomes

$$\begin{pmatrix} 0.104\ldots & 2.5 & 1.770\ldots \\ 1.25 & 0 & 1.25 \\ 1.770\ldots & 2.5 & 0.104\ldots \end{pmatrix}.$$

Now $-T(x;\lambda)$ can be rewritten as $Bx - (1+g)Ax$, where

$$B \equiv \begin{pmatrix} 0.1 & 0.4 & 0.3 \\ 0.2 & 0.1 & 0.2 \\ 0.3 & 0.4 & 0.1 \end{pmatrix}, \; A \equiv \begin{pmatrix} 0.4 & 0 & 0 \\ 0 & 0.4 & 0 \\ 0 & 0 & 0.4 \end{pmatrix},$$

and $0 \leq g < 0.693\ldots$. Therefore, interpreting B as the material output coefficient matrix, A as the input matrix, and g as the rate of steady balanced growth, our Leontief output equation $Bx - (1+g)Ax = d$ has a nonnegative solution for any nonnegative final demand vector $d \in R^3_+$ when the rate of balanced growth is not less than 0 and smaller than $0.693\ldots$. Each process may be identified as an *industry,* not by its output, but by its input!

We proceed to make a nonlinear extension of the above. Let $T(x;\lambda) \equiv \lambda x - A(x)$, where $A(x)$ is a nonlinear map from $X = R^n$ to R^n with $n \geq 2$. We revise the assumptions 2.1. and 2.2, making the latter stronger, and add one more assumption on $A(x)$.

Assumption 3.1. There exists an $x^* \gg 0$ in R^n and a scalar $\lambda > 0$ such that $T(x^*;\lambda) \gg 0$.

Assumption 3.2.

(i) The map $A(x)$ is indecomposable, i.e., for any $x \in R^n$ and any two nonempty disjoint subsets, I and J of index set $N \equiv \{1, 2, \ldots, n\}$ such that $I \cup J = N$, at least one element of $A(x+y)$ in J is greater than the corresponding element of $A(x)$ when $y_i > 0$ for $i \in I$ and $y_j = 0$ for $j \in J$.

(ii) $A(x)$ is isotone on R^n_+, i.e., $A(x) \geq A(y)$ if $x \geq y$ for $x, y \in R^n_+$;

Assumption 3.3.

(i) $A(x)$ is continuous on R^n, and $A(0) = 0$;

(ii) $A(x)$ is quasi-convex and positively homogeneous of degree one, i.e., $A(kx) = kA(x)$ for any positive scalar k.

It should be noted that the assumption 3.2 makes the following assumption 3.2* valid, which is a parameterized version of the assumption 2.2*.

Assumption 3.2*. If $T(x;\lambda) > 0$ for $x \in R^n_+$ and for an arbitrary positive scalar λ, then $x \gg 0$.

In the linear case, when A is an $n \times n$ nonnegative indecomposable matrix, all the assumptions 3.1, 3.2, and 3.3 are satisfied.

First we present

Proposition 3.1. *Given the assumptions 3.1, 3.2, and 3.3, there exists a* λ^* *such that (i) if* $\lambda > \lambda^*$, $T(x;\lambda) = d$ *has a unique solution* $x \gg 0$ *for any* $d > 0$, *and (ii)* $T(x;\lambda^*) = 0$ *has a nonzero solution* $x^* \gg 0$, *which is unique up to scalar multiplication.*

Proof. These are the results contained in Morishima and Fujimoto[7] and Fujimoto[4], thus the proof is omitted. □

Proposition 3.2. *Given the assumptions 3.1, 3.2, and 3.3, there exists a* $\lambda^o > 0$ *such that when* $\lambda^o < \lambda < \lambda^*$, $-T(x;\lambda)$ *satisfies the assumptions 2.1, 2.2*, 2.3, and 2.4.*

Proof. When $\lambda < \lambda^*$, we have $-T(x^*;\lambda) \gg 0$ for $x^* \gg 0$, thus $-T(x;\lambda)$ satisfies the assumption 2.1. Suppose the assumption 2.2* does not hold for $-T(x;\lambda)$ in any interval like $\lambda^o < \lambda < \lambda^*$. This implies that corresponding to a series $\{\lambda_i\}$, which is increasing and converging to λ^*, there exists a nonnegative nonzero vector series $\{x_i\}$ such that at least one element of each x_i is zero, and $-T(x_i;\lambda_i) > 0$. Since the dimension is finite and $A(x)$ is continuous, we can find $x^{**} \in R^n_+$ and $x^{**} \notin \text{int}R^n_+$ such that $-T(x^{**};\lambda^*) \geq 0$. If this were $T(x^{**};\lambda^*) = 0$, it would be a contradiction to the assumption 3.2 of indecomposability, or the uniqueness of x^*, a solution to $T(x;\lambda^*) = 0$. If the inequality were $T(x^{**};\lambda^*) < 0$, this would imply by the indecomposability the existence of $x \gg 0$ such that $T(x;\lambda^*) \ll 0$. This is a contradiction to the property of λ^* that $\lambda^* = \sup\{\lambda |\, \lambda \geq 0, T(x;\lambda) \leq 0 \text{ for some } x > 0\}$. (See Morishima and Fujimoto[7].) We can choose λ^o to be positive near λ^*.

Therefore, we have found that there exists an interval of λ, $\lambda^o < \lambda < \lambda^*$, within which all the assumptions 2.1, 2.2*, 2.3, and 2.4 are satisfied for $-T(x;\lambda)$. □

Proposition 3.3. *If* $A(x)$ *is linear, i.e., a nonnegative indecomposable* n *by* n *matrix A, there exists a* $\lambda^o > 0$ *such that when* $\lambda^o < \lambda < \lambda^*$, $-(\lambda x - Ax)$ *has a strictly positive inverse. (*λ^* *is the Frobenius root of A.)*

Proof. This proposition is evident because Proposition 3.2 shows Propositions 2.2 and 2.3 are valid. □

4. Examples

As in a simple Leontief model, when the spaces X and Y are the same and the Euclidean space of dimension n with $n \geq 2$, and the transformation is $T(x) \equiv x - Ax$, where A is an $n \times n$ real nonnegative matrix, the assumptions 2.2, 2.3, and 2.4 are naturally satisfied. Thus, all we have to examine is the assumption 2.1, the condition of *productiveness*. This case is well known and the reader is referred to Berman and Plemmons[1].

The case in which $T(x) \equiv x - A(x)$, where $A(x)$ is in general a nonlinear isotone mapping, and the spaces can be complete lattices of any dimension, is treated in Fujimoto[4]. The assumption 2.2 is again satisfied by the form of mapping itself. Proposition 2.1 can be obtained without using the assumption 2.3 also by the very form of the map $T(x) \equiv x - A(x)$.

Now we take up the linear case in which the spaces X and Y are the same and the Euclidean space of dimension n with $n \geq 2$, and the map is given as $Tx \equiv Bx - Ax$, where A and B are both $n \times n$ real matrices. This can be interpreted as a Leontief model with joint production. The assumptions 2.3 and 2.4 are clearly satisfied. Thus we have to check the assumptions 2.1 and 2.2 to be able to secure our propositions. Several numerical examples are given in Fujimoto et al.[5]. One of the interesting cases is when T is represented as

$$T \equiv \begin{pmatrix} 1 & -a & 1 \\ 1 & 1 & -a \\ -a & 1 & 1 \end{pmatrix}.$$

Then T has the inverse

$$\begin{pmatrix} \frac{-1}{a^2-a-2} & \frac{-1}{a^2-a-2} & \frac{-a+1}{a^2-a-2} \\ \frac{-a+1}{a^2-a-2} & \frac{-1}{a^2-a-2} & \frac{-1}{a^2-a-2} \\ \frac{-1}{a^2-a-2} & \frac{-a+1}{a^2-a-2} & \frac{-1}{a^2-a-2} \end{pmatrix},$$

when $a \neq -1$ or $a \neq 2$. This inverse is nonzero and nonnegative if $1 \leq a < 2$. It is not difficult to see the assumptions 2.1 and 2.2 are satisfied for T when $1 \leq a < 2$. And it is easy to generalize the above example T to a real square matrix having an odd number of columns and rows. When a real matrix has an odd number ($2k + 1$ with $k \geq 1$) of columns and rows, its diagonal elements are all unity, and in each row, shifting from the diagonal element to the right, $(-a)$ and 1 appear alternately(jumping back to the first element at the rightmost one), then it is inverse-positive when $1 \leq a < (1 + \frac{1}{k})$. (Each element of the inverse contains a fraction whose denominator is $(ka^2 - a - (k+1))$, while its numerator is either $((k-1)a - k)$ or $(-a+1)$.) In this case also, it is not hard to see the assumptions 2.1 and 2.2 are satisfied. Therefore, for example

$$\begin{pmatrix} 1 & -1 & 1 & -1 & 1 \\ 1 & 1 & -1 & 1 & -1 \\ -1 & 1 & 1 & -1 & 1 \\ 1 & -1 & 1 & 1 & -1 \\ -1 & 1 & -1 & 1 & 1 \end{pmatrix}$$

has its inverse with all entries nonnegative.

To raise an example from nonlinear cases, we consider the following

$$T(x) \equiv \begin{pmatrix} 1 & -b_{12} & 1 \\ 1 & 1 & -b_{23} \\ -1 & 1 & 1 \end{pmatrix} x,$$

where b_{12} and b_{23} now depend upon x as

$$b_{12} \equiv \begin{cases} 2 - \log_e(x_2 + 1)/x_2, & \text{when } x_2 > 0 \\ 1, & \text{when} x_2 \leq 0 \end{cases},$$

and

$$b_{23} \equiv \begin{cases} 1 + x_3/(x_2 + 1), & \text{when } x_2 > 0 \\ 1 + x_3, & \text{when} x_2 \leq 0 \end{cases}.$$

The first component function

$$T_1(x) \equiv \begin{cases} x_1 - (2x_2 - \log_e(x_2 + 1)) + x_3, & \text{when } x_2 > 0 \\ x_1 - x_2 + x_3, & \text{when} x_2 \leq 0 \end{cases},$$

as well as the second function

$$T_2(x) \equiv \begin{cases} x_1 + x_2 - (x_3 + x_3^2/(x_2 + 1)), & \text{when } x_2 > 0 \\ x_1 + x_2 - (x_3 + x_3^2), & \text{when} x_2 \leq 0 \end{cases}$$

are concave. The assumption 2.1 is satisfied by $e \equiv (1, 1, 1)'$, and it is not difficult to see the assumption 2.2 is also fulfilled. Thus, our propositions are applicable. The input coefficient b_{12} implies the decreasing returns to scale of process 2 because b_{12} is increasing in x_2 when x_2 is positive, while b_{23} means the external economies rendered by process 2 on process 3 because the input coefficient b_{23} is decreasing in x_2 when x_2 is positive.

5. Comments

It should be noted that we do not depend upon the theory of determinants at all, and the nonnegative invertibility of matrices have been established in an elementary way. Moreover nonlinear generalizations have come out naturally.

Positive invertibility in section 3 does not seem to need the special form, $x - A(x)$, but can be extended to the case, $B(x) - A(x)$, or even more generally to the case $T(x)$ with some additional conditions on T. In the linear case, there is no problem in extending along this line. Therefore, T in section 4 of examples, when made negative $-T$,

$$-T \equiv \begin{pmatrix} -1 & a & -1 \\ -1 & -1 & a \\ a & -1 & -1 \end{pmatrix},$$

has its inverse, when $a = 4$,

$$\begin{pmatrix} 0.1 & 0.1 & 0.3 \\ 0.3 & 0.1 & 0.1 \\ 0.1 & 0.3 & 0.1 \end{pmatrix}.$$

Exceptionally with this 3×3 case, $-T$ is inverse-positive for any $a > 2$, converging to the 0 matrix as the parameter value a approaches infinity.

References

[1] Berman, A., Plemmons, R.J.: Nonnegative Matrices in the Mathematical Sciences. New York, Academic Press 1979

[2] Bidard, C.: Prix, reproduction, rareté. Paris: Dunod 1991 (Chapter 10, p.113)

[3] Chander, P.: A simple proof of the nonsubstitution theorem. Quarterly Journal of Economics **88**, 698-701 (1974)

[4] Fujimoto, T.: Nonlinear Leontief models in abstract spaces. Journal of Mathematical Economics **15**, 151-156 (1986)

[5] Fujimoto, T., Silva, J., Villar, A.: An input-output model with joint production, mimeograph. Department of Fundamental Analysis of Economics, University of Alicante 2002

[6] Herrero, C., Villar, A.: A characterization of economies with the non-substitution property. Economics Letters **26**, 147-152 (1988)

[7] Morishima, M., Fujimoto, T.: The Frobenius theorem, its Solow-Samuelson extension and the Kuhn-Tucker theorem. Journal of Mathematical Economics **1**, 199-205 (1974)

[8] Schefold, B.: Multiple product techniques with properties of single product systems. Zeitschrift für Nationalökonomie **38**, 29-53 (1978)

Adv. Math. Econ. 5, 65–96 (2003)

Advances in
MATHEMATICAL
ECONOMICS

Implicit functions and diffeomorphisms without C^1*

Leonid Hurwicz[1] **and Marcel K. Richter**[2]

[1] Department of Economics, University of Minnesota, 271 19th Avenue South, Minneapolis, MN 55455, USA
(e-mail: hurwicz@tc.umn.edu)
[2] Department of Economics, University of Minnesota, 271 19th Avenue South, Minneapolis, MN 55455, USA
(e-mail: richter@econ.umn.edu)

Received: August 1, 2002
Revised: August 22, 2002

JEL classification: C600, C650

Mathematics Subject Classification (2000): 26B10, 58C15

Abstract. We prove implicit and inverse function theorems for non-C^1 functions, and characterize non-C^1 diffeomorphisms.

Key words: Implicit function theorem, inverse function theorem, diffeomorphism, non-C^1 analysis

1. Introduction

With K representing either the real field $\mathbb{R}$ or the complex field $\mathbb{C}$, suppose $f(\bar{x}, \bar{y}) = 0$ for some function $f\colon X \times Y \to K^k$, where $X \subseteqq K^n$ and $Y \subseteqq K^k$. We seek a solution, an "implicit function," ϕ on some neighborhood of $\bar{x}$ in which $f(x, \phi(x)) = 0$ holds.

Cauchy, working with complex scalars, originally assumed that f was analytic [10], and Dini's formulation for real scalars assumed that f was C^1 [15]. Similar smoothness assumptions have formed the backbone of most proofs

* We are indebted to Professor Wayne H. Richter, University of Minnesota; Professor Hülya Eraslan, University of Pennsylvania; and Nevzat Eren, University of Minnesota, for valuable comments on an earlier version.

since then. But what if f is only differentiable, rather than C^1 — and only at $(\bar{x}, \bar{y})$? It will be shown that a solution ϕ still exists.

When f is differentiable at $(\bar{x}, \bar{y})$ and continuous in a neighborhood of $(\bar{x}, \bar{y})$, Halkin [23] showed that an implicit function does exist locally. (He allowed X to be an arbitrary normed linear space.)

When $g\colon X \to K^n$ is differentiable at $\bar{x}$ and has a nonzero Jacobian determinant, and when it is also locally continuous, Halkin also proved that a right inverse exists. Further, when the Jacobian determinant of g is nonzero in a neighborhood, Radulescu and Radulescu [36] showed that g is a local diffeomorphism.

When X as well as Y are finite dimensional, we extend Halkin's point-result for the implicit function theorem by weakening the continuity assumption on f. We do this with an alternative application of Brouwer's fixed point theorem. We also apply the fixed point theorem with Brouwer's theorem on invariance of domain to give an alternative proof of Radulescu and Radulescu's result, without explicit mention of degree theory.

Our first two theorems weaken the C^1 assumption — to either differentiability at $(\bar{x}, \bar{y})$, or to partial differentiability with respect to y at $(\bar{x}, \bar{y})$, together with some continuity conditions on f. In the former case we obtain differentiability of all solutions at $\bar{x}$; in the latter case we obtain continuity of all solutions at $\bar{x}$. A third theorem combines the two frameworks. Solutions need not be unique. However, when f is differentiable in a neighborhood of $(\bar{x}, \bar{y})$ and f_y is surjective there, then there is a unique solution and it is differentiable.

Theorem 1 yields a General Inverse Function Theorem (Theorem 4), which in turn characterizes diffeomorphisms without continuous differentiability.

Because implicit and inverse function theorems play a role in several parts of mathematics, there are many applications.

2. General implicit function theorems

Let K be the real field $\mathbb{R}$ or the complex field $\mathbb{C}$, and let $\|\cdot\|$ be any norm on the finite dimensional Banach space K^q over K. For any $z \in K^q$ and real $\varepsilon \geqq 0$, we denote by $B_\varepsilon^{K^q}(z)$ the closed ball in K^q centered at z with $\|\cdot\|$-radius ε. When we talk of continuity of functions on subsets $A \subseteqq K^q$, we mean continuity with respect to the relative topology induced on A from K^q. We denote by $\mathrm{Int}(A)$ the interior of any subset A of K^q.

We will need to bound the norm on K^{n+k} in terms of the norms on K^n and K^k. We note that the values $\|(x,0)\|$ of elements $(x,0) \in K^{n+k}$ define a norm on K^n. Since all norms on a finite dimensional real vector space are equivalent, there exists a real $\beta_1 > 1$ such that:

$$\|(x,0)\| \leqq \beta_1 \|x\| \qquad \text{for all } x \in K^n. \tag{1a}$$

Analogously, there exists a real $\beta_2 > 1$ such that:

$$\|(0, y)\| \leqq \beta_2 \|y\| \qquad \text{for all } y \in K^k. \tag{1b}$$

Consequently: for all real ζ and all $x \in K^n$ and $y \in K^k$,

$$\|y\| \leqq \zeta\|x\| \quad \Rightarrow \quad \|(x, y)\| \leqq (\beta_1 + \beta_2\zeta)\|x\|. \tag{2}$$

If X is a subset of K^q, we denote by id_X the identity function on X.

For $\bar{u} \in U \subseteqq K^n$ and $f\colon U \to K^k$, we say f is Lipschitz at $\bar{u}$ (for $\|u\| \leqq \mu$) if there is some real $\mu > 0$ and some real $\gamma \geqq 0$ such that:

$$\|f(u + \bar{u}) - f(\bar{u})\| \leqq \gamma\|u\| \qquad \text{for all } u \in K^n \text{ with } \|u\| \leqq \mu. \tag{3}$$

By differentiability we will always mean Fréchet differentiability. Because we do not always assume that the domain of f is open, derivatives, when they exist, need not be unique; and in that case $f'(\bar{u})$ will denote any derivative.

If $f\colon X \to Y$ for some nonempty set $X \subseteqq K^q$ and some set $Y \subseteqq K^q$, then we say that $f\colon X \to f(X)$ is a diffeomorphism if f is injective, $f(X)$ is open, and both f and its inverse $f^{-1}\colon f(X) \to X$ are differentiable. And we say that f is a local diffeomorphism at $\bar{u} \in X$, if there is some nonempty open neighborhood $U \subseteqq X$ of $\bar{x}$ such that $f\colon U \to f(U)$ is a diffeomorphism. If a diffeomorphism and its inverse are r times differentiable, then we say it is a D^r diffeomorphism.

While the classical proofs of implicit or inverse function theorems often use contraction mapping theorems, we will use Brouwer's fixed point theorem for finite dimensional Banach spaces.[1] (Usually stated for $K = \mathbb{R}$, it holds for $K = \mathbb{C}$ as well, since every finite dimensional complex Banach space is isomorphic (hence homeomorphic) to $\mathbb{C}^n$ for some n, which in turn is homeomorphic to $\mathbb{R}^{2n}$.)

We will also use the finite dimensional Banach space version of another theorem of Brouwer, which follows immediately from the result [6] for real scalars by the isomorphism and homeomorphism just mentioned:

> (Brouwer's Theorem on Invariance of Domain.) If A is an open subset of K^n and $g\colon A \to K^n$ is a continuous injection, then $g(A)$ is an open subset of K^n, hence g is an open mapping, so g^{-1} is continuous and g is a homeomorphism of A onto $g(A)$.

Our first theorem assumes differentiability of f at a point, and yields differentiability of solutions at the point (in parts (a) and (b)).

---

[1] An earlier application of Brouwer's fixed point theorem to an implicit function theorem was made in [23]. We were unaware of Halkin's result when we wrote the initial versions of this paper [25]. See the details, p. 94.

Theorem 1. Differentiable implicit function theorem. *With $K = \mathbb{R}$ or $K = \mathbb{C}$, let X be a subset of K^n, let Y be a subset of K^k, and let $U \subseteqq X \times Y$. Let $\bar{u} = (\bar{x}, \bar{y}) \in U$, where $\bar{x}$ is a limit point of X, and $\bar{y} \in Y$.*[2] *Suppose that $f : U \to K^k$ and:*

$$f(\bar{x}, \bar{y}) = 0; \tag{4a}$$

$$f(\,\cdot\,,\,\cdot\,) \text{ is differentiable at } (\bar{x}, \bar{y}), \tag{4b}$$
$$\text{with partial derivatives } f_x(\bar{x}, \bar{y}) \text{ and } f_y(\bar{x}, \bar{y});$$

$$f_y(\bar{x}, \bar{y}) \text{ is surjective ; i.e.,} \tag{4c}$$

$$\det\left(\begin{bmatrix} \frac{\partial f^1(\bar{x},\bar{y})}{\partial y_1} & \cdots & \frac{\partial f^1(\bar{x},\bar{y})}{\partial y_k} \\ \vdots & & \vdots \\ \frac{\partial f^k(\bar{x},\bar{y})}{\partial y_1} & \cdots & \frac{\partial f^k(\bar{z})}{\partial y_k} \end{bmatrix}\right) \neq 0.$$

Suppose that, for some $\gamma > \|(f_y(\bar{x}, \bar{y}))^{-1} f_x(\bar{x}, \bar{y})\|$:[3] *for all x, y,*

$$x \in X \text{ and } y \in B^{K^k}_{\gamma\|x\|}(\bar{y}) \quad \Rightarrow \tag{4d}$$
$$(x, y) \in U \text{ and } f(x,\,\cdot\,) \text{ is defined and continuous on } B^{K^k}_{\gamma\|x\|}(\bar{y}).$$

For any $\xi > 0$, let $X_\xi = B^{K^n}_\xi(\bar{x}) \cap X$.
Then:

a) There exists a real $\bar{\delta} > 0$ and a function $\phi \in \prod_{x \in X_{\bar{\delta}}} B^{K^k}_{\gamma\|x\|}(\bar{y})$ such that: for all $x \in X_{\bar{\delta}}$,

$$f(x, \phi(x)) = 0 \tag{5a}$$
$$\phi(\bar{x}) = \bar{y}. \tag{5b}$$

b) The $\bar{\delta}$ in part (a) can be chosen so that every function $\phi \in \prod_{x \in X_{\bar{\delta}}} B^{K^k}_{\gamma\|x\|}(\bar{y})$ satisfying (5) is differentiable at $\bar{x}$.

c) For every $\xi > 0$, all functions $\phi \in \prod_{x \in X_\xi} B^{K^k}_{\gamma\|x\|}(\bar{y})$ satisfying (5) that are differentiable at $\bar{x}$ have the same derivative $\phi'(\bar{x})$:

$$\phi'(\bar{x}) = -\,(f_y(\bar{x}, \bar{y}))^{-1} f_x(\bar{x}, \bar{y}). \tag{6}$$

---

[2] In textbook cases, X, Y, and U are open sets. Here we are not assuming that X even contains an open set; it could, for example, contain just $\bar{x}$ and a sequence of points $x^i \to \bar{x}$.

[3] By the norm of a linear transformation L, we mean $\|L\| = \max\{\|Lx\| : \|x\| = 1\}$.

d) If U is open, then the derivatives $f_x(\bar{x},\bar{y})$, $f_y(\bar{x},\bar{y})$, and $\phi'(\bar{x})$ are uniquely determined.

e) If X is open and if (4b,4c,4d) are replaced by these stronger conditions:

$$f \text{ is differentiable on } X \times Y \tag{4e}$$

$$f_y \text{ is surjective on } X \times Y, \tag{4f}$$

then an open neighborhood $X_0 \times Y_0 \subseteqq X \times Y$ of $(\bar{x},\bar{y})$ can be chosen so that:

i) there is a unique function $\phi\colon X_0 \to Y_0$ satisfying (5) for $x \in X_0$;

ii) ϕ is differentiable on X_0, with:

$$\phi'(x) = -\left(f_y(x,\phi(x))\right)^{-1} f_x(x,\phi(x)). \tag{7}$$

f) (Classical) *If X is open, and if f is C^r on $X \times Y$ for some $r \geqq 1$, then there is an open neighborhood $X_0 \times Y_0$ of $(\bar{x},\bar{y})$ and a unique function $\phi\colon X_0 \to Y_0$ satisfying (5), and ϕ is C^r on X_0.*

Remark 1 We cannot totally eliminate the continuity hypothesis (4d). Consider the function $f\colon \mathbb{R}^2 \to \mathbb{R}$ defined by:

$$f(x,y) = \begin{cases} x+y, & \text{if } x+y \neq 0 \\ x^2+y^2, & otherwise. \end{cases} \tag{8}$$

The local continuity hypothesis of Theorem 1 fails, since $f(x,\cdot)$ is discontinuous at all $(x,y) \neq (0,0)$ with $x+y=0$. So for every $x \neq 0$, the function $f(x,\cdot)$ is not continuous even on $B^{K^k}_{\|x\|}(0)$. Yet all other hypotheses hold: $f(0,0)=0$, f is differentiable at $(0,0)$, and $f_y(0,0)=1$. Here one cannot solve for y as a function of $x \neq 0$, since no (x,y) in $\mathbb{R}^2$, other than $(0,0)$, satisfies $f(x,y)=0$.

It is tempting, therefore, to try and prove the theorem assuming that, for every x, the function $f(x,\cdot)$ is continuous on $B^{K^k}_{\|x\|}(\bar{y})$. But our next example shows that is not possible: Theorem 1's requirement on γ is "tight." First we define the function $g\colon \mathbb{R} \to \mathbb{R}$ by:

$$g(x) = -(x+x^3), \tag{9}$$

and then we define the function $f\colon \mathbb{R}^2 \to \mathbb{R}$ by:

$$f(x,y) = \begin{cases} x-g^{-1}(y), & \text{if } x-g^{-1}(y) \neq 0 \\ x^2+y^2, & \text{otherwise.} \end{cases} \tag{10}$$

For each x, the function $f(x,\cdot)$ is continuous on $B^{K^k}_{\|x\|}(0)$; that is clear from the figure indicating typical level curves.

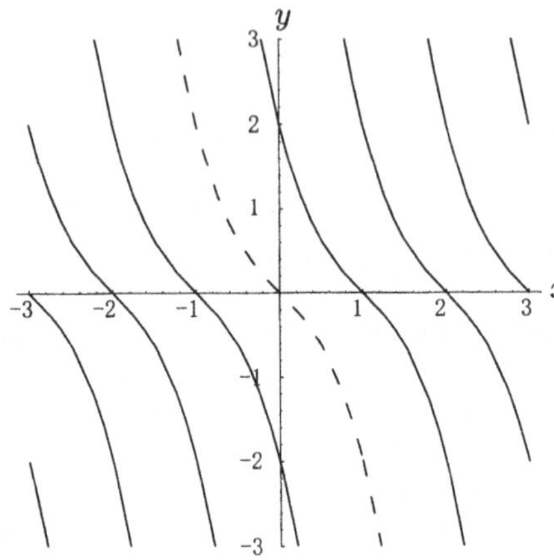

Figure 1

Here the dashed level curve through the origin is the locus of the only discontinuities of f. Even though all other assumptions of the theorem hold, again there is no implicit function solving for y as a function of $x \neq 0$: $f(x, y) = 0$ only for $(x, y) = (0, 0)$.

Remark 2 Theorem 1 replaces the classical C^1 hypotheses of Cauchy's (complex) theorem and Dini's (real) theorem[4] by the weaker assumptions (4b,4d). Its conclusions (a,b) are weaker than the classical versions, as the implicit functions ϕ are not necessarily unique or C^1. The conclusion (e) is part of the classical Implicit Function Theorems.

Remark 3 Simple examples show that, under the hypotheses of parts (a), (b), and (c), the function ϕ need not be unique.

Proof of Theorem 1. Without loss of generality, we simplify the discussion by assuming that:

$$\bar{x} = 0 \in K^n \text{ and } \bar{y} = 0 \in K^k. \tag{11}$$

(a) (Existence) To prove the existence assertions (5), it suffices to prove:

$$\exists \bar{\delta}_{\ \bar{\delta}>0}\ \forall x_{\ \|x\| \leqq \bar{\delta}\ \&\ x \in X}\ \exists y_{\ \|y\| \leqq \gamma \|x\|}\ \ (x, y) \in U\ \ \&\ \ f(x, y) = 0. \tag{12}$$

For (12) guarantees for each $x \in X_{\bar{\delta}}$ the existence of a point $y \in B^{K^k}_{\gamma \|x\|}(\bar{y})$ such that:

$$f(x, y) = 0 \tag{13a}$$

$$x = 0 \Rightarrow y = 0 \tag{13b}$$

$$\|y\| \leqq \gamma \|x\|. \tag{13c}$$

[4] See the historical notes in Section 4.

Thus we can define a function ϕ on $X_{\bar{\delta}}$ taking values $\phi(x) = y \in B^{K^k}_{\gamma\|x\|}(\bar{y})$ with:

$$\phi(\bar{x}) = \bar{y} \tag{14}$$

and with the properties (5).[5]

We break our proof of (12) into parts corresponding to its quantifiers.

i) Writing:

$$M = \|(f_y(0,0))^{-1} f_x(0,0)\|, \tag{15}$$

we define:

$$\alpha = \gamma - M, \tag{16}$$

which is positive by hypothesis.

ii) The choice of $\bar{\delta}$ is delicate, and will involve several steps. First note that, since f is differentiable at $(0,0)$ with partial derivatives $f_x(0,0)$ and $f_y(0,0)$, we have: for all $(x,y) \in K^n \times K^k$ with $(x,y) \in U$,

$$f(x,y) = f(0,0) + f_x(0,0) \cdot x + f_y(0,0) \cdot y + r(x,y), \tag{17}$$

where $r(x,\,\cdot\,)$ is $o(\|(x,y)\|)$, i.e.,

$$\frac{\|r(x,y)\|}{\|(x,y)\|} \to 0 \qquad \text{for all } (x,y) \in U \text{ with } 0 \neq \|(x,y)\| \to 0. \tag{18}$$

Note that the function $r(x,\,\cdot\,)$ determined by (17) is defined and continuous on $B^{K^k}_{\gamma\|x\|}(\bar{y})$ for each $x \in X$, by hypothesis (4d).

By (4c) we can define $T = (f_y(0,0))^{-1}$.

Now for any given $x \in X$, satisfying (12) means finding $y \in B^{K^k}_{\gamma\|x\|}(\bar{y})$ with $f(x,y) = 0$, a task that can be stated in each of these equivalent forms: $(x,y) \in X \times Y$ and

$$f(x,y) = 0 \tag{19a}$$

$$f_x(0,0) \cdot x + f_y(0,0) \cdot y + r(x,y) = 0 \quad \text{(by (17) and (4a))} \tag{19b}$$

$$r(x,y) = -f_x(0,0) \cdot x - f_y(0,0) \cdot y \quad \text{(rearranging)} \tag{19c}$$

$$Tr(x,y) = -Tf_x(0,0) \cdot x - Tf_y(0,0) \cdot y \quad \text{(applying } T\text{)} \tag{19d}$$

$$Tr(x,y) = -Tf_x(0,0) \cdot x - y \quad \text{(by definition of } T\text{)} \tag{19e}$$

$$y = -Tr(x,y) - Tf_x(0,0) \cdot x \quad \text{(rearranging).} \tag{19f}$$

[5] The Axiom of Choice is not needed for this. For example, we could use the lexicographic ordering based on $y_1, \ldots, y_k$ to choose among multiple candidates y with minimum norm in each of the compact sets $(f(\bar{x}+x,\,\cdot\,))^{-1}(0) \cap B_{\gamma\delta}(\bar{y})$.

In (iii) below, for every $x \in X$ near 0, we will apply a fixed point theorem to find a solution y of (19f), hence of (19a). To prepare for the fixed point theorem, we will bound the two terms on the right hand side of (19f).

Note that:

$$\text{The equivalences (19) hold for all } x \in X \text{ and } y \text{ with } \|y\| \leqq \gamma\|x\|, \text{ since then (4d) implies that } f(x,y) \text{ and thus } r(x,y) \text{ are defined.} \tag{20}$$

ii.1) To bound the first term on the right in (19f), we note from (18) that:

$$\frac{\|Tr(x,y)\|}{\|(x,y)\|} \to 0 \quad \text{for all } (x,y) \in U \text{ with } 0 \neq \|(x,y)\| \to 0, \tag{21}$$

so by (20):

$$\begin{aligned} \forall\varepsilon_{\ \varepsilon>0}\ \exists\delta_{\ \delta>0}\ \forall(x,y)_{\ \|(x,y)\|\leqq\delta\ \&\ x\in X\ \&\ y\in B^{K^k}_{\gamma\|x\|}(\bar{y})} \\ \|Tr(x,y)\| &\leqq \varepsilon\|(x,y)\| \\ &\leqq \varepsilon(\beta_1+\beta_2\gamma)\|x\| \qquad \text{(by (2)).} \end{aligned} \tag{22}$$

Then for $\varepsilon = \alpha/(\beta_1+\beta_2\gamma)$, we have:

$$\exists\delta_{\ \delta>0}\ \forall(x,y)_{\ \|(x,y)\|\leqq\delta\ \&\ x\in X\ \&\ y\in B^{K^k}_{\gamma\|x\|}(\bar{y})}\ \|Tr(x,y)\| \leqq \alpha\|x\|. \tag{23}$$

Let δ be any such number. Defining:

$$\bar{\delta} = \delta/(\beta_1+\beta_2\gamma), \tag{24}$$

we see that: if $\|x\| \leqq \bar{\delta}$ and $\|y\| \leqq \gamma\|x\|$ then,

$$\begin{aligned} \|(x,y)\| &\leqq (\beta_1+\beta_2\gamma)\|x\| \qquad \text{(by (2))} \\ &\leqq (\beta_1+\beta_2\gamma)\bar{\delta} \\ &= \delta. \end{aligned} \tag{25}$$

From (20), (25), (23), and (4d) we have:

$$\|x\| \leqq \bar{\delta}\ \ \&\ \ \|y\| \leqq \gamma\|x\| \quad \Rightarrow \quad (x,y) \in U\ \ \&\ \ \|Tr(x,y)\| \leqq \alpha\|x\|. \tag{26}$$

ii.2) To bound the second term in the right hand side of (19f), note that by the definition of M: for all $x \in K^n$,

$$\|Tf_x(\bar{x},\bar{y}) \cdot x\| \leqq M\|x\|. \tag{27}$$

iii) To complete the proof of (12) we begin by picking any $x \in X$ with $\|x\| \leqq \bar{\delta}$. We must find a $y \in Y$ with $\|y\| \leqq \gamma\|x\|$ and $f(x,y) = 0$. This last equality means finding a y equating the left and right sides of (19f). Representing the right side through the function $F\colon B^{K^k}_{\gamma\|x\|}(\bar{y}) \to K^k$ defined by:

$$F(y) = -Tr(x,y) - Tf_x(\bar{x},\bar{y}) \cdot x, \tag{28}$$

we see that we are seeking a fixed point of $F(\,\cdot\,)$.

Clearly F is continuous in y because $r(x,\,\cdot\,)$ is (by (17) and (4d)). And for all y with $\|y\| \leqq \gamma\|x\|$:

$$\begin{aligned}\|F(y)\| &= \|-Tr(x,y) - Tf_x(\bar{x},\bar{y}) \cdot x\| \\ &\leqq \|Tr(x,y)\| + \|Tf_x(\bar{x},\bar{y}) \cdot x\| \\ &\leqq (\alpha + M)\|x\| \qquad \text{(by (26) and (27))} \\ &= \gamma\|x\| \qquad \text{(by (16)).}\end{aligned} \tag{29}$$

Thus F is a continuous function that carries the ball $B^{K^k}_{\gamma\|x\|}(\bar{y}) \subseteqq Y$ into itself. By Brouwer's Fixed Point Theorem, F must have a fixed point $F(y) = y \in B^{K^k}_{\gamma\|x\|}(\bar{y})$, and therefore (by the equivalence of (19f) and (19a)) we have $f(x,y) = 0$.

Using our definition (14), this allows us to define an implicit function ϕ on $B^{K^n}_{\bar{\delta}}(\bar{x})$ satisfying (5).

Note that our construction of ϕ was based on a particular choice of $f_x(\bar{x},\bar{y})$ and $f_y(\bar{x},\bar{y})$.[6] Since U is not necessarily open, these are not necessarily uniquely determined, and a different choice may give a different definition of ϕ.

(b) (Differentiability at $\bar{x}$) To prove that $\bar{\delta}$ can be chosen in part (a) so that every implicit function $\phi \in \prod_{x\in X_{\bar{\delta}}} B^{K^k}_{\gamma\|x\|}(\bar{y})$ is differentiable, we first show that $\bar{\delta}$ can be chosen so that any implicit function is Lipschitz at $\bar{x} = 0$.

b.i) The particular implicit function constructed in our proof of part (a) has the Lipschitz property at $\bar{x} = 0$, as we see from (13c) and our method of construction in (a.iii). By itself, however, this does not prove that all functions ϕ satisfying (5) have the Lipschitz property. To show that *every* function ϕ satisfying part (a) of the Theorem has the Lipschitz property at $\bar{x} = 0$ for x in some neighborhood of $\bar{x}$, we must show that:

$$\begin{gathered}\exists\delta_{\ \delta>0}\ \forall\phi_{\ \phi\in \mathrm{IM}(\delta)}\ \exists\varepsilon_{\ \varepsilon>0}\ \exists\eta_{\ \eta>0}\ \forall x_{\ x\in X_\delta\ \&\ \|x\|\leqq\varepsilon} \\ \|\phi(x)\| \leqq \eta\|x\|,\end{gathered} \tag{30}$$

[6] Cf. the definition of M in (15), the definition of γ at the beginning of part (a,i) of the proof, the definition of α in (16), the definition of $r(x,y)$ in (17), the definition of δ in (22), the definition of $\bar{\delta}$ in (24), the definition of δ in part (ii.2) preceding (27), and the definition of F in (28).

where we write $\phi \in \mathrm{IM}(\delta)$ to indicate that $\phi \in \prod_{x \in X_\delta} B^{K^k}_{\gamma\|x\|}(\bar{y})$ is an implicit mapping, i.e., $\phi(x) \in B^{K^k}_{\gamma\|x\|}(\bar{y})$ and $\phi(0) = 0$ and $f(x, \phi(x)) = 0$ for all $x \in X_\delta$.

Suppose that (30) fails, so:

$$\begin{gathered} \forall\delta_{\ \delta>0}\ \exists\phi_{\ \phi\in\mathrm{IM}(\delta)}\ \forall\varepsilon_{\ \varepsilon>0}\ \forall\eta_{\ \eta>0}\ \exists x_{\ x\in X_\delta\ \&\ \|x\|\leqq\varepsilon} \\ \|\phi(x)\| > \eta\|x\|. \end{gathered} \tag{31}$$

We will obtain a contradiction.

For each $i \in \mathbb{N}$, pick a positive real δ^i, with $\delta^i \searrow 0$ as $i \to \infty$. Then by (31) we can choose, for each $i \in \mathbb{N}$:

$$\text{a function } \phi^i \in \mathrm{IM}(\delta^i), \text{ so } \phi^i \in \Pi_{x \in X_{\delta^i}} B^{K^k}_{\gamma\|x\|}(\bar{y}) \tag{32a}$$

$$\text{a positive } \varepsilon^i \text{ with } \varepsilon^i \searrow 0 \text{ for } i \to \infty \tag{32b}$$

$$\text{a positive } \eta^i \text{ with } \eta^i \to \infty \text{ for } i \to \infty \tag{32c}$$

$$x^i \in X_{\delta^i}\ \&\ 0 < \|x^i\| < \varepsilon^i \text{ with } \|\phi^i(x^i)\| > \eta^i\|x^i\|. \tag{32d}$$

It follows from (32a,32d) that $\eta^i\|x^i\| < \|\phi^i(x^i)\| \leqq \gamma\|x^i\|$, so $\eta^i < \gamma$. This contradiction of (32c), completes our proof that every solution ϕ of part (a) must be Lipschitz at $\bar{x}$.

b.ii) To prove Theorem 1(b), let ϕ be any solution of part (a), and consider any value $f_x(0,0)$, still assuming that $\bar{x} = 0$ and $\bar{y} = 0$. It suffices to show that:

$$\frac{\|\phi(x) - \phi(0) + (f_y(0,0))^{-1} f_x(0,0)x\|}{\|x\|} \xrightarrow[x \in X\ \&\ \|x\|\to 0]{} 0. \tag{33}$$

From (b.i) we know that ϕ is Lipschitz at $\bar{x} = 0$ in some X-neighborhood of $\bar{x}$, say with Lipschitz constant $\tilde{\gamma}$.

Since f is differentiable at $(0,0)$,

$$f(x, \phi(x)) = f(0,0) + f_x(0,0)x + f_y(0,0)\phi(x) + o(\|(x, \phi(x))\|) \tag{34}$$

for all $x \in X$, so by (11) and (5) we have:

$$\frac{\|f_x(0,0)x + f_y(0,0)\phi(x)\|}{\|(x, \phi(x))\|} \xrightarrow[x \in X\ \&\ \|(x,\phi(x))\|\to 0]{} 0. \tag{35}$$

With β_1 and β_2 as in (1), we have:

$$\begin{aligned} \|(x, \phi(x))\| &\leqq \|(x,0)\| + \|(0, \phi(x))\| \\ &\leqq \beta_1\|x\| + \beta_2\|\phi(x)\| \\ &\leqq (\beta_1 + \beta_2\tilde{\gamma})\|x\|, \end{aligned} \tag{36}$$

for all $x \in X$ near 0. So (35) implies:

$$\frac{\|f_x(0,0)x + f_y(0,0)\phi(x)\|}{\|x\|} \xrightarrow[x\in X \ \& \ \|x\|\to 0]{} 0. \tag{37}$$

Then because $(f_y(0,0))^{-1}$ is linear,

$$\frac{\|(f_y(0,0))^{-1}f_x(0,0)x + \phi(x)\|}{\|x\|} \xrightarrow[x\in X \ \& \ \|x\|\to 0]{} 0, \tag{38}$$

and this reformulation of (33) verifies that ϕ is differentiable at $\bar{x} = 0$.

(c) The claim of Theorem 1(c) is clear from our proof of part (b.ii); in particular, the proof of (6) follows from (38).

(d) Here X and Y can be taken as open, so the derivatives $f_y(\bar{x}, \bar{y})$ and $f_x(\bar{x}, \bar{y})$ are uniquely determined, and then parts (b) and (c) guarantee that, for all $\xi, \eta > 0$, and all (X_ξ, Y_η, ϕ) satisfying part (a) of Theorem 1, the derivative $\phi'(\bar{x})$ exists, with the unique value given by (6).

(e) Let $\widetilde{X}_{\bar{\delta}}$ and $\widetilde{\phi}\colon X_{\bar{\delta}} \to B^{K^k}_{\gamma\|x\|}(\bar{y}) \subseteqq Y$ have the properties of $X_{\bar{\delta}}$, and ϕ of part (a) of Theorem 1. Since we can now assume that both X and Y are open, we can assume that $\widetilde{X}_{\bar{\delta}}$ is also open.

(e.i) (Uniqueness on X_0) We will pick an open $X_0 \subseteqq \widetilde{X}_{\bar{\delta}}$ and an open $Y_0 \subseteqq Y$ in order to find a unique implicit function ϕ. First define $G\colon X \times Y \to X \times K^k$ by:

$$G(x, y) = (x, f(x, y)). \tag{39}$$

By hypothesis (4e), G is differentiable on $X \times Y$, and from hypothesis (4f) we see that the derivative $G'(x, y)$ is surjective at all $(x, y) \in X \times Y$. Then parts (a) and (c.ii) of Theorem 4 (the General Inverse Function Theorem)[7] imply there is an open neighborhood $X_\diamond \times Y_\diamond \subseteqq X \times Y$ of $G(\bar{x}, \bar{y}) = (\bar{x}, 0)$ such that $G\restriction_{X_\diamond \times Y_\diamond}$ is a diffeomorphism to the open neighborhood $G(X_\diamond \times Y_\diamond)$ of $G(\bar{x}, \bar{y}) = (\bar{x}, 0)$. So there is an open neighborhood $X_0 \subseteqq \widetilde{X}_{\bar{\delta}}$ of $\bar{x}$ such that, for every $x \in X_0$ there is a unique $y \in Y_\diamond$ with $G(x, y) = (x, 0)$, i.e., with $f(x, y) = 0$. For such x we also have $f(x, \widetilde{\phi}(x)) = 0$, so $y = \widetilde{\phi}(x)$. Thus $\phi = \widetilde{\phi}\restriction_{X_0}$ is the unique implicit function from X_0 into $Y_\diamond$. Therefore defining $Y_0 = Y_\diamond$ completes the proof of (e.i).

(e.ii) (Differentiability on X_0) Let X_0, Y_0, and $\phi\colon X_0 \to Y_0$ be as in part (e.i) of the Theorem, and let $\tilde{x}$ be any element of X_0. Define $\tilde{y} = \phi(\tilde{x})$. Then:

[7] There is no circularity in our reasoning here. Our proof below of parts (a) and (c) of Theorem 4 will depend only on parts (a), (b), and (c) (but not (e)) of the present theorem.

$$f(\tilde{x},\tilde{y}) = 0 \qquad \text{(by definition of } \phi \text{ and } \tilde{y}\text{);} \tag{40a}$$

$$f(x,\cdot\,) \text{ is continuous on } Y, \text{ for all } x \in X \qquad \text{(by (4d));} \tag{40b}$$

$$f(\,\cdot\,,\cdot\,) \text{ is differentiable at } (\tilde{x},\tilde{y}) \qquad \text{(by (4e));} \tag{40c}$$

$$f_y(\tilde{x},\tilde{y}) \text{ is surjective} \qquad \text{(by (4f)).} \tag{40d}$$

Therefore the assumptions (4a,4b,4c,4d) of Theorem 1 are satisfied with $(\tilde{x},\tilde{y})$ replacing $(\bar{x},\bar{y})$. So the conclusions (a), (d), and (e.i) of Theorem 1 hold with $(\tilde{x},\tilde{y})$ replacing $(\bar{x},\bar{y})$. Thus $\phi\colon X_0 \to Y_0$ is differentiable at $\tilde{x}$ and (7) holds at $x = \tilde{x}$. Since $\tilde{x}$ was an arbitrary element of X_0, the claim (e.ii) follows.

(f) (Classical) The uniqueness follows from part (e.ii); and the C^r property follows from (7) when f is C^r. Of course the uniqueness and C^r assertions, as well as formula (7) are part of the classical Implicit Function Theorem. □

Even as simple a function as this:

$$f(x,y) = y - (x)^{\frac{1}{3}} \tag{41}$$

fails to satisfy the differentiability hypothesis of the preceding theorem. And, while it clearly admits the unique implicit function $\phi(x) = (x)^{\frac{1}{3}}$, that solution is not differentiable at 0. Indeed, the same would be true of any function

$$f(x,y) = y - \psi(x), \tag{42}$$

with ψ nondifferentiable. Such examples motivate our next implicit function theorem, which in the spirit of Goursat [19] has a weaker differentiability hypothesis (no differentiability with respect to x) but a stronger continuity hypothesis (joint in x and y). The conclusion of our second theorem will be weaker than our first, asserting merely continuity of solutions rather than differentiability at the initial point.

Theorem 2. Continuous implicit function theorem. *With $K = \mathbb{R}$ or $K = \mathbb{C}$, let X be an open subset of K^n and let Y be an open subset of K^k. Suppose that $(\bar{x},\bar{y}) \in X \times Y$ and $f\colon X \times Y \to K^k$. Suppose that:*

$$f(\bar{x},\bar{y}) = 0; \tag{43a}$$

$$f(\,\cdot\,,\cdot\,) \text{ is continuous on } X \times Y; \tag{43b}$$

$$f(\bar{x},\cdot\,) \text{ is differentiable at } \bar{y}; \tag{43c}$$

$$f_y(\bar{x},\bar{y}) \text{ is surjective ; i.e.,} \tag{43d}$$

$$\det\left(\begin{bmatrix} \dfrac{\partial f^1(\bar{x},\bar{y})}{\partial y_1} & \cdots & \dfrac{\partial f^1(\bar{x},\bar{y})}{\partial y_k} \\ \vdots & & \vdots \\ \dfrac{\partial f^k(\bar{x},\bar{y})}{\partial y_1} & \cdots & \dfrac{\partial f^k(\bar{x},\bar{y})}{\partial y_k} \end{bmatrix}\right) \neq 0.$$

Then:

a) *There exist open subsets $X_0 \subseteqq X$ and $Y_0 \subseteqq Y$ with $(\bar{x}, \bar{y}) \in X_0 \times Y_0$, and there exists a function $\phi\colon X_0 \to Y_0$ such that:*

$$f(x, \phi(x)) = 0 \qquad \text{for all } x \in X_0 \tag{44a}$$
$$\phi(\bar{x}) = \bar{y}. \tag{44b}$$

b) *The set $X_0 \times Y_0$ in (a) can be chosen so that there exists at least one function $\phi\colon X_0 \to Y_0$ satisfying (44), and so that every such function is continuous at $\bar{x}$.*

c) (Goursat.) *If assumption (43c) is replaced by this stronger condition:*

$$f(x, \cdot\,) \text{ is } C^1 \text{ on } Y \qquad \text{for all } x \in X, \tag{43e}$$

then there is a neighborhood $X_0 \times Y_0$ of $(\bar{x}, \bar{y})$ and a unique function $\phi\colon X_0 \to Y_0$ satisfying (44a,44b), and ϕ is continuous on X_0.

Remark 4 As W. H. Young showed [39], the existence result (a) for the case $k = 1$ only requires that f be continuous separately in x and y, rather than jointly. See the historical remarks below.

Remark 5 Theorem 2(a,b) replaces the C^1 hypothesis of Goursat's result [19][8] by the weaker assumptions (43b,43c). Its conclusions in (a) and (b) are weaker than in Goursat [19], as the implicit functions ϕ are not necessarily unique or continuous. Here is an example for $K = \mathbb{R}$ in which the uniqueness and continuity conclusions of part (c) cannot be obtained under the non-C^1 hypotheses of parts (a) and (b): Let $f(x, y) = x - g(y)$, where

$$g(y) = \begin{cases} (y/2) + y^2 \sin(1/y), & \text{if } y \neq 0 \\ 0, & \text{otherwise.} \end{cases} \tag{45}$$

Proof of Theorem 2. Without loss of generality, we assume that:

$$\bar{x} = 0 \qquad \text{and} \qquad \bar{y} = 0. \tag{46}$$

(a) (Existence) Without loss of generality, we assume that:[9]

$$f_y(0, 0) = \mathrm{id}_{K^k}\,. \tag{47}$$

This is justified since $f_y(0, 0)$ is invertible (43d), so we could define the function $g(x, y) = f(x, (f_y(0, 0))^{-1} y)$, which would have all the properties (43)

[8] See the historical notes in Section 4., and the comparisons with results of W. H. Young and of Bliss.

[9] Recall that id_{K^k} is the identity function on K^k. Cf. p. 67.

with $g_y(0,0)$ the identity on K^k, and g would admit an implicit function ϕ if and only if f admits an implicit function $\tilde{\phi} = (f_y(0,0))^{-1}\phi$:

$$g(x,\phi(x)) = 0 \Leftrightarrow f(x,\tilde{\phi}(x)) = 0. \tag{48}$$

To prove existence of an implicit function under (47), then, define $F\colon X \times Y \to K^k$ by:

$$F(x,y) = y - f(x,y). \tag{49}$$

By hypothesis (43c) and by (46) and (47),

$$\begin{aligned} f(0,y) &= f(0,0) + f_y(0,0)y + R(y) \\ &= y + R(y) \end{aligned} \tag{50}$$

for some function $R(y)$ of y that is $o(\|y\|)$:

$$\frac{\|R(y)\|}{\|y\|} \xrightarrow[0 \neq \|y\| \to 0]{} 0. \tag{51}$$

Then:

$$\begin{aligned} \frac{\|F(0,y)\|}{\|y\|} &= \frac{\|y - f(0,y)\|}{\|y\|} \qquad \text{(by (49) and (50))} \\ &= \frac{\|R(y)\|}{\|y\|} \xrightarrow[0 \neq \|y\| \to 0]{} 0 \qquad \text{(by (51))}, \end{aligned} \tag{52}$$

so there exists a $\tilde{\gamma} > 0$ such that $B_{\tilde{\gamma}}(\bar{y}) \subseteqq Y$ and:[10]

$$\forall \gamma_{\ 0<\gamma\leqq\tilde{\gamma}} \ \forall y_{\ \|y\|\leqq\gamma} \ \|F(0,y)\| < \gamma. \tag{53}$$

We next show that:

$$\forall \gamma_{\ 0<\gamma\leqq\tilde{\gamma}} \ \exists \varepsilon_{\gamma\ 0<\varepsilon_\gamma\leqq\bar{\varepsilon}} \ \forall \varepsilon_{\ 0\leqq\varepsilon\leqq\varepsilon_\gamma} \ \forall x_{\ \|x\|\leqq\varepsilon} \ \forall y_{\ \|y\|\leqq\gamma} \ \|F(x,y)\| < \gamma. \tag{54}$$

We begin by fixing any positive $\gamma \leqq \tilde{\gamma}$. Then let $\bar{\varepsilon} > 0$ be small enough that $B_{\bar{\varepsilon}}(\bar{x}) \subseteqq X$, and define the correspondence $C\colon \{\varepsilon \in K : 0 \leqq \varepsilon \leqq \bar{\varepsilon}\} \twoheadrightarrow X \times Y$ by:

$$C(\varepsilon) = \{(x,y) \colon \|x\| \leqq \varepsilon \ \& \ \|y\| \leqq \gamma\}, \tag{55}$$

and the corresponding maximum function:

[10] Our proof of (53) is based on the differentiability hypothesis (43c). Alternatively, we could obtain (53) by simply assuming that $F(\bar{x},\cdot)$ has a Lipschitz constant less than 1 at $y = 0$. That property is a weakened version of the hypotheses of Goursat's lemma [19] (p. 185, §1, for $k = 1, n = 1$, and p. 188, §3, for arbitrary k), which sets the stage for his application of what is now known as the Contraction Mapping Theorem. But our continuity proof in part (b) will still use the differentiability hypothesis (43c).

$$M(\varepsilon) = \max\{\|F(x,y)\| \colon (x,y) \in C(\varepsilon)\}. \tag{56}$$

Since F is continuous by hypothesis (43b), and since $C(\varepsilon)$ is clearly a continuous correspondence,[11] it follows from the Maximum Theorem[12] that $M(\,\cdot\,)$ is a continuous function. So (54) follows from (53).

Thus for every $\gamma \leqq \tilde{\gamma}$ and every x with $\|x\| \leqq \varepsilon_\gamma$, the function $F(x,\,\cdot\,)$ carries $B_\gamma(0)$ into $B_\gamma(0)$. It is also continuous by (43b), so by Brouwer's Fixed Point Theorem there exists a $y \in B_\gamma(0)$ such that

$$F(x,y) = y, \tag{57}$$

i.e.,

$$f(x,y) = 0. \tag{58}$$

If $x \neq 0$, then we define $\phi(x) = y$, choosing any such y;[13] if $x = 0$ we define $\phi(0) = 0$. Then (44) holds, with $X_0 = B_\varepsilon(0)$ and $Y_0 = B_\gamma(0)$.

We have thus shown:

> For every $\gamma > 0$ with $\gamma \leqq \tilde{\gamma}$, and every $\varepsilon \geqq 0$ with $\varepsilon \leqq \varepsilon_\gamma$, there exists a function $\phi\colon X_\gamma \to Y_\gamma$ such that (44) holds with $X_0 = B_\varepsilon(0)$ and $Y_0 = B_\gamma(0)$. (59)

(b) (Continuity) We now prove that the set $X_0 \times Y_0$ containing $(\bar{x}, \bar{y})$ in part (a) of Theorem 2 can be chosen so that not only does there exist a function $\phi\colon X_0 \to Y_0$ satisfying (44), but every such function is continuous at $x = 0$.

We can choose $\delta > 0$ with $\delta \leqq \tilde{\gamma}$ such that:

$$\|R(y)\| < \|y\| \qquad \text{for all } y \in B_\delta(0) \qquad \text{(using (51))}, \tag{60a}$$

and by (50) we can also ensure that:

$$y \neq 0 \Rightarrow f(0,y) \neq 0 \qquad \text{for all } y \in B_\delta(0). \tag{60b}$$

Defining $\gamma = \delta$, $X_0 = B_{\varepsilon_\gamma}(0)$ and $Y_0 = B_\gamma(0)$, it follows from (59) that there exists a function $\phi\colon X_0 \to Y_0$ satisfying (44).

Now let ϕ be any function from X_0 into Y_0 satisfying (44). Suppose that ϕ is not continuous at 0. Then there is a sequence $x_1, x_2, \ldots$ converging to 0, but with $\phi(x_i)$ not converging to $\phi(0) = 0$. So there is a subsequence, whose elements we again denote by x_i, for which

$$\phi(x_i) \xrightarrow[i\to\infty]{} \tilde{y} \neq 0. \tag{61}$$

---

[11] I.e., both upper and lower hemicontinuous. Cf. Berge [1], [2], where "semicontinuity" is used for what we are calling "hemicontinuity."

[12] Cf. [12], p. 889, Remark, [13], p. 19, 1.8(4), [1], p. 122, [2], p. 116.

[13] As in footnote 5, the Axiom of Choice is not needed.

By (60b) that implies:

$$f(0,\tilde{y}) \neq 0, \tag{62}$$

so by the continuity hypothesis (43b) we have $f(x,y) \neq 0$ for all (x,y) in some neighborhood of $(0,\tilde{y})$. But defining $y_j = \phi(x_j)$ we also have $f(x_j,y_j) = 0$ by the solution property (44a), and $(x_j,y_j) \xrightarrow[i\to\infty]{} (0,\tilde{y})$ by (61). This contradiction completes the proof of continuity.

(c) (Uniqueness and continuity) If $f(x,\cdot\,)$ is C^1 on Y for all $x \in X$, then the local uniqueness and continuity of the implicit function are part of Goursat's result [19]. □

Remark 6 While some of the hypotheses of the Continuous and the Differentiable Implicit Function Theorems are the same, in total they are noncomparable. Parts (a) and (b) of Theorem 1 have stronger differentiability hypotheses than parts (a) and (b) of Theorem 2, but weaker continuity properties. While the function f in (41) satisfies the hypotheses of Theorem 2(a,b) but not those of the Theorem 1(a,b), the reverse is true for the function

$$f(x,y) = \begin{cases} y + x^2 + y^2, & \text{if } x \text{ is rational} \\ y, & \text{otherwise.} \end{cases} \tag{63}$$

Here $f(x,y)$ is not continuous in x at $(0,y)$ for $y \neq 0$, so it violates (43b) for $(\bar{x},\bar{y}) = (0,0)$; but it is differentiable at $(0,0)$, and satisfies the other hypotheses of Theorem 1(a,b).

We have obtained implicit functions under two different hypotheses about the independent variable x. In Theorem 1(a,b), the function $f(x,y)$ was differentiable jointly in x and y at $(\bar{x},\bar{y})$; and we concluded that the implicit function $\phi(x)$ was differentiable at $\bar{x}$. In Theorem 2(a,b), the function $f(x,y)$ was only differentiable with respect to y at $(\bar{x},\bar{y})$, and $f(x,y)$ was continuous with respect to x at $\bar{x}$; and we concluded that the implicit function $\phi(x)$ was continuous at $\bar{x}$. Now we show that these two results can be combined, allowing two types of independent variables — those satisfying differentiability hypotheses yield differentiability conclusions for the implicit function, and those satisfying continuity hypothesis yield continuity conclusions. We are interested in solving $f(v,w,y) = 0$ for y as a function ϕ of (v,w): $f(v,w,\phi(v,w)) = 0$. The role played by x in Theorem 1 is played here by v, and the role of x in Theorem 2 is played here by w.

Theorem 3. General implicit function theorem. *With $K = \mathbb{R}$ or $K = \mathbb{C}$, let V, W, and Y be subsets of K^m, K^p, and K^k, respectively, where $m + p > 0$; we allow $m = 0$ or $p = 0$, in which case V or W, respectively, is empty. Suppose that $(\bar{v},\bar{w},\bar{y}) \in V \times W \times Y$, that $\bar{v}$ is a limit point of V, and that W and Y are open. Let $f\colon V \times W \times Y \to K^k$. Suppose that:*

$$f(\bar{v}, \bar{w}, \bar{y}) = 0; \tag{64a}$$

$$f(v, \cdot, \cdot) \text{ is continuous on } W \times Y, \text{ for all } v \in V; \tag{64b}$$

$$f(\cdot, \bar{w}, \cdot) \text{ is differentiable at } (\bar{v}, \bar{y}); \tag{64c}$$

$$f_y(\bar{v}, \bar{w}, \bar{y}) \text{ is surjective ; i.e.,} \tag{64d}$$

$$\det\left(\begin{bmatrix} \frac{\partial f^1(\bar{v}, \bar{w}, \bar{y})}{\partial y_1} & \cdots & \frac{\partial f^1(\bar{v}, \bar{w}, \bar{y})}{\partial y_k} \\ \vdots & & \vdots \\ \frac{\partial f^k(\bar{v}, \bar{w}, \bar{y})}{\partial y_1} & \cdots & \frac{\partial f^k(\bar{v}, \bar{w}, \bar{y})}{\partial y_k} \end{bmatrix}\right) \neq 0.$$

For any $\xi, \eta > 0$, let $V_\xi = \mathrm{Int}(B_\xi(\bar{v})) \cap V$ and $Y_\eta = \mathrm{Int}(B_\eta(\bar{y}))$. Then:

a) There exist real $\xi, \eta > 0$ with $Y_\eta \subseteqq Y$ such that, for every $v \in V_\xi$ there exists an open neighborhood W_v of $\bar{w}$ in W, and a function $\phi(v, \cdot)\colon W_v \times Y_\eta$ such that: for all $w \in W_v$,

$$f(v, w, \phi(v, w)) = 0 \tag{65a}$$

$$\phi(\bar{v}, \bar{w}) = \bar{y}. \tag{65b}$$

b) The ξ and η in part (a) can be chosen so that every function ϕ satisfying part (a) also satisfies:

$$\phi(\bar{v}, \cdot) \text{ is continuous at } \bar{w} \tag{66a}$$

$$\phi(\cdot, \bar{w}) \text{ is differentiable at } \bar{v}. \tag{66b}$$

c) For every $\xi, \eta > 0$, all functions $\phi(\cdot, \bar{w})\colon V_\xi \to Y_\eta$ satisfying (65) that are differentiable at $\bar{v}$ have derivatives $\phi_v(\bar{v}, \bar{w})$, determined by the derivatives $f_v(\bar{v}, \bar{w}, \bar{y})$ as follows:

i) for any value of $f_v(\bar{v}, \bar{w}, \bar{y})$, a value of $\phi_v(\bar{x})$ is:

$$\phi_v(\bar{v}, \bar{w}) = -\left(f_y(\bar{v}, \bar{w}, \bar{y})\right)^{-1} f_v(\bar{v}, \bar{w}, \bar{y}); \tag{67a}$$

ii) for any value of $\phi_v(\bar{v}, \bar{w})$, a value of $f_v(\bar{v}, \bar{w}, \bar{y})$ is:

$$f_v(\bar{v}, \bar{w}, \bar{y}) = -f_y(\bar{v}, \bar{w}, \bar{y})\phi_v(\bar{v}, \bar{w}). \tag{67b}$$

d) If $f(\cdot, \cdot, \cdot)$ is continuous on $V \times W \times Y$, then there is a neighborhood $V_0 \times W_0 \times Y_0$ and a function ϕ satisfying (65) that is continuous at $(\bar{v}, \bar{w})$.

e) (Goursat [19], [20].) *If $f(\cdot, \cdot, \cdot)$ is continuous on $V \times W \times Y$, and if $f(v, w, \cdot)$ is C^1 on Y for all $(v, w) \in V \times W$, then there is a neighborhood $V_1 \times W_1 \times Y_1$ of $(\bar{v}, \bar{w}, \bar{y})$ and a unique function $\phi\colon V_1 \times W_1 \to Y_1$ satisfying (65a,65b), and ϕ is continuous on $V_1 \times W_1$.*

Remark 7 This theorem is a generalization of the Differentiable and the Continuous Implicit Function Theorems: when w is absent, the hypotheses and conclusions are those of the Differentiable Implicit Function Theorem, and when v is absent, they are those of the Continuous Implicit Function Theorem.

Proof of Theorem 3. Without loss of generality, we simplify by assuming that:

$$\bar{v} = 0 \quad \text{and} \quad \bar{w} = 0 \quad \text{and} \quad \bar{y} = 0. \tag{68}$$

Also without loss of generality, we assume that:

$$f_y(0,0,0) = \mathrm{id}_{K^k}. \tag{69}$$

This is justified since $f_y(0,0,0)$ is invertible (64d), so we could define the function $g(v,w,y) = f(v,w,(f_y(0,0,0))^{-1}y)$, which would have all the properties (64) with $g_y(0,0,0)$ the identity on K^k, and g would admit an implicit function ϕ if and only if f admits an implicit function $\tilde{\phi} = (f_y(0,0,0))^{-1}\phi$:

$$g(v,w,\phi(v,w)) = 0 \Leftrightarrow f(v,w,\tilde{\phi}(v,w)) = 0. \tag{70}$$

(a) (Existence) Define $F\colon V \times W \times Y$ by:

$$F(v,w,y) = y - f(v,w,y). \tag{71}$$

By hypothesis (64c),

$$\begin{aligned} f(v,0,y) &= f(0,0,0) + f_v(0,0,0)v + f_y(0,0,0)y + R(v,y) \\ &= y + f_v(0,0,0)v + R(v,y) \qquad \text{(by (69), (64a) and (68))} \end{aligned} \tag{72}$$

for some function $R(v,y)$ of (v,y) that is $o(\|(v,y)\|)$:

$$\frac{R(v,y)}{\|(v,y)\|} \xrightarrow[0 \neq \|(v,y)\| \to 0]{} 0. \tag{73}$$

By (71) and (72) we have:

$$\begin{aligned} \|F(v,0,y)\| &= \|y - f(v,0,y)\| \\ &= \|R(v,y) + f_v(0,0,0)v\| \\ &\leqq \frac{\|R(v,y)\|}{\|(v,y)\|}\|(v,y)\| + \|f_v(0,0,0)\| \, \|v\|, \end{aligned} \tag{74}$$

so by (73) there exists a $\gamma > 0$ such that, for all $\|(v,y)\| < \gamma$ we have

$$\forall (v,y)_{\ \|(v,y)\| \leqq \gamma} \ \|F(v,0,y)\| < \frac{1}{2}\gamma + \|f_v(0,0,0)\| \, \|v\|. \tag{75}$$

Now define:

$$\varepsilon_0 = \begin{cases} \dfrac{\gamma}{2\|f_v(0,0,0)\|}, & \text{if } \|f_v(0,0,0)\| \neq 0 \\ \\ \frac{1}{2}\gamma, & \text{otherwise,} \end{cases} \tag{76}$$

so $\varepsilon_0 > 0$. Now let

$$\varepsilon = \min\{\varepsilon_0, \gamma\}. \tag{77}$$

Then:

$$\forall v_{\ \|v\|<\varepsilon}\ \forall y_{\ \|(v,y)\|\leqq\gamma}\ \|F(v,0,y)\| < \frac{1}{2}\gamma + \frac{1}{2}\gamma = \gamma. \tag{78}$$

Without loss of generality, we may suppose that the norm on $V \times Y$ is the maximum norm, and then (78) and (77) imply :

$$\forall v_{\ \|v\|<\varepsilon}\ \forall y_{\ \|y\|\leqq\gamma}\ \|F(v,0,y)\| < \gamma. \tag{79}$$

Next we verify that this inequality holds not only for $w = 0$, but for all w near 0. Since $F(v,0,y)$ is defined for $\|v\| < \varepsilon\ \ \&\ \ \|y\| \leqq \gamma$, openness of W ensures there is some $\bar{\delta} > 0$ for which $F(v,w,y)$ is defined for all $\|v\| < \varepsilon\ \ \&$ $\|w\| \leqq \bar{\delta}\ \ \&\ \ \|y\| \leqq \gamma$. Then paralleling the proof of Theorem 2,[14] we define:

$$C(\delta) = \{(w,y) : \|w\| \leqq \delta\ \ \&\ \ \|y\| \leqq \gamma\}, \tag{80}$$

for all δ with $0 \leqq \delta \leqq \bar{\delta}$. Clearly $C(\,\cdot\,)$ is a continuous correspondence, so by the continuity hypothesis (64b) and the Maximum Theorem, for every v with $\|v\| < \varepsilon$, the function

$$M_v(\delta) = \max\{\|F(v,w,y)\| \colon \|w\| \leqq \delta\ \ \&\ \ \|y\| \leqq \gamma\} \tag{81}$$

is continuous for $0 \leqq \delta \leqq \bar{\delta}$. So by (79) for every v with $\|v\| < \varepsilon$, there exists a $\delta_v > 0$ with $\delta_v \leqq \bar{\delta}$ such that:

$$\forall v_{\ \|v\|<\varepsilon}\ \forall w_{\ \|w\|\leqq\delta_v}\ \forall y_{\ \|y\|<\gamma}\ \|F(v,w,y)\| \leqq \gamma. \tag{82}$$

Thus for every v with $\|v\| < \varepsilon$ and every w with $\|w\| \leqq \delta_v$, the function $F(v,w,\,\cdot\,)$ carries $B_\gamma(0)$ into $B_\gamma(0)$. It is also continuous by (64b), so by Brouwer's Fixed Point Theorem there exists a $y \in B_\gamma(0)$ such that

$$F(v,w,y) = y, \tag{83}$$

i.e.,

$$f(v,w,y) = 0. \tag{84}$$

[14] Cf. (54), (55), and (56).

If $w \neq 0$, then we define $\phi(v,w) = y$, for any such y;[15] if $w = 0$ and $v = 0$, we define $\phi(0,0) = 0$. Then (65) holds with $V_0 = B_\varepsilon(0)$, $W_v = B_{\delta_v}(0)$ and $Y_v = B_\gamma(0)$.

(b) (Continuity with respect to w) To prove part (66a), we note that the function defined by $\tilde{f}(w,y) = f(0,w,y)$ satisfies the hypotheses of the Continuous Implicit Function Theorem, and so by part (b) of that theorem we immediately obtain property (66a).

(Differentiability with respect to v.) To prove (66b), we note that the function $\hat{f}$ defined by $\hat{f}(v,y) = f(v,0,y)$ satisfies the hypotheses of the Differentiable Implicit Function theorem, and so by part (b) of that theorem we immediately obtain property (66b).

(c) Cf. Theorem 1(c).

(d) (Continuity with respect to (v,w)) Part (c) follows immediately from part (b) of the Continuous Implicit Function Theorem, since if we define $x = (v,w)$, then all the hypotheses of that theorem apply.

(e) (Uniqueness and local continuity) See part (c) of Theorem 2. □

3. A general inverse function theorem

As usual, one can derive an inverse function theorem from an implicit function theorem.

Theorem 4. General inverse function theorem. *Let $K = \mathbb{R}$ or $K = \mathbb{C}$, and let X be an open subset of K^n, and let $g\colon X \longrightarrow K^n$ be continuous on X and differentiable at $\bar{x} \in X$. Let $\bar{y} = g(\bar{x})$. Suppose $g'(\bar{x})$ is surjective, i.e., $\operatorname{rank}(g'(\bar{x})) = n$. Then:*

a) There exists an open neighborhood Y of $\bar{y}$ and a function $h\colon Y \to X$ such that:

$$g \circ h = \mathrm{id}_Y \tag{85a}$$

$$h(\bar{y}) = \bar{x}. \tag{85b}$$

b) For every open neighborhood Y of $\bar{y}$, every function h satisfying (85) is injective, and is differentiable at $g(\bar{x})$ with:

$$h'(g(\bar{x})) = (g'(\bar{x}))^{-1}. \tag{86}$$

c) If the differentiability of g at $\bar{x}$ is strengthened to differentiability on X, and surjectivity of $g'(\bar{x})$ is strengthened to surjectivity of g' on X, then:

[15] As in footnote 5, the Axiom of Choice is not needed.

i) For every open neighborhood Y of $\bar{y}$, every function $h\colon Y \to X$ satisfying (85) is differentiable on Y and h' is surjective on Y.

ii) An open neighborhood X_0 of $\bar{x}$ can be chosen so that $g\restriction_{X_0}$ is a diffeomorphism from X_0 to the open neighborhood $g(X_0)$ of $\bar{y}$; in particular, $g\restriction_{X_0}$ has a unique inverse h on $g(X_0)$, and:

$$h'(g(x)) = (g'(x))^{-1} \qquad \text{for all } x \in X_0. \tag{87}$$

d) (Classical) *If g is C^r on X for some $r \geqq 1$, then there is an open neighborhood X_0 of $\bar{x}$ such that $g\restriction_{X_0}$ is a C^r-diffeomorphism from X_0 to the open set $g(X_0)$.*

Proof We will prove the General Inverse Function Theorem from the Differentiable Implicit Function Theorem, reversing the roles of x and y. Define

$$f(x, y) = g(x) - y \tag{88}$$

for all $x \in X$ and $y \in K^n$, so f is differentiable at $(\bar{x}, \bar{y})$.

(a) (Existence) To prove part (a), note that because $f_x(\bar{x}, \bar{y}) = g'(\bar{x})$ is surjective, Theorem 1(a) implies that there exists an open neighborhood Y of $\bar{y} = g(\bar{x})$ and a function $\psi\colon Y \to X$ such that:

$$f(\psi(y), y) = 0 \qquad \text{for all } y \in Y \tag{89a}$$

$$\psi(\bar{y}) = \bar{x}. \tag{89b}$$

So in view of (88), $h = \psi$ satisfies (85); thus h is a right inverse on Y of g.

(b) (Injectivity) Let Y and h be as in part (a). Then h is injective: if $h(y) = h(\tilde{y})$ then $y = g(h(y)) = g(h(\tilde{y})) = \tilde{y}$ by (85a).

(Differentiability at $g(\bar{x})$.) Theorem 1(b) implies that any ψ satisfying (89) is differentiable at $\bar{y} = g(\bar{x})$. Then by (85) the Chain Rule implies that $g'(h(\bar{y}))h'(\bar{y}) =$ the identity on Y, so $h = \psi$ also satisfies (86).

(c.i) (Differentiability on Y) Assume now that $g'(x)$ exists and is surjective for all $x \in X$, let Y be an open neighborhood of $\bar{y}$, and let $h\colon Y \to X$ satisfy (85) as guaranteed by Theorem 4(a). For any $\tilde{y} \in Y$ define $\tilde{x} = h(\tilde{y}) \in X$, so $\tilde{y} = g(\tilde{x})$. Then by Theorem 4(b) h is differentiable at $\tilde{y}$, and $h'(\tilde{y}) = (g'(\tilde{x}))^{-1}$.

(c.ii) (Diffeomorphism) By Theorem 4(a) there exists an open neighborhood Y of $\bar{y}$ and a function $h\colon Y \to X$ satisfying (85). Then h is injective on Y (by (b)) and continuous on Y (by (c.i) since g is differentiable on X). So it follows from Brouwer's theorem on Invariance of Domain that h is a homeomorphism from Y to the open neighborhood $h(Y)$ of $\bar{x}$. Thus $X_0 = h(Y)$ is an open neighborhood of $\bar{x}$, and $g\restriction_{X_0}\colon X_0 \to g(X_0) = g(h(Y)) = Y$, as a left inverse of h, is also a homeomorphism; so $g\restriction_{X_0}$ and h are inverse homeomorphisms between X_0 and Y. By (c.i) the inverse h is differentiable, so $g\restriction_{X_0}$ is a diffeomorphism and (87) holds.

(c) (Classical) The uniqueness follows from part (c.ii); and the C^r property follows from (87) when the homeomorphism $g \restriction X_0$ is C^r. Of course the uniqueness and C^r assertions, as well as formula (87), are contained in the classical Inverse Function Theorem. □

Diffeomorphism Corollary. *Let $K = \mathbb{R}$ or $K = \mathbb{C}$, let X be an open subset of K^n, let $g\colon X \to K^n$, and let the integer $r \geqq 1$. Then g is a local D^r diffeomorphism*[16] *at $\bar{x} \in X$ if and only if there is some neighborhood of $\bar{x}$ on which g is D^r and g' is surjective.*

Proof The necessity of local D^r differentiability and surjectivity is obvious. The sufficiency is contained in part (c.ii) of the General Inverse Function Theorem when $r = 1$. For $r > 1$ the proof follows by induction, paralleling the usual proofs of the classical Inverse Function Theorem.[17] □

Remark 8

a) There are examples[18] in which no inverse h in part (a) of the General Inverse Function Theorem is unique or continuous. In such examples, with f as defined by (88), none of the functions ϕ in Theorems 1(a,b), 2(a,b), and 3(a,b) can be unique or continuous.

b) The continuity hypothesis on g cannot be dropped altogether, as shown by instances based on example (8) above.

c) The uniqueness and differentiability properties that follow from the additional hypotheses in part (c) are new results in the real case $K = \mathbb{R}$. When $K = \mathbb{C}$, however, the local differentiability assumption in (c) is only apparently weaker than the classical assumptions for $K = \mathbb{R}$, since for $K = \mathbb{C}$ the local differentiability hypothesis of (c)[19] implies, by Goursat's Theorem [11], p. 100 on analytic functions, that g is actually analytic — in which case the classical hypotheses and conclusions of part (d) hold.

d) As an example of a diffeomorphism that is not C^1, yet satisfies our hypotheses in part (c.ii) of Theorem 4 and the Corollary, consider:

$$g(x) = \begin{cases} 2x + x^2 \sin(1/x), & \text{for } -1 < x < 1 \\ 0, & \text{for } x = 0. \end{cases} \tag{90}$$

e) The theorems have useful applications, since the classical Implicit Function Theorem is imbedded in many parts of mathematics, including differentiable manifolds and optimization theory. The weaker hypotheses of the theorems above admit the possibility of extending existing results to these and other areas.

---

16 Cf. page 67.
17 Cf. [14], p. 272, (10.2.5).
18 E.g. (45) or [14], p. 273, Problem 2.
19 But not those of parts (a) and (b).

For example, replacing the classical Implicit Function Theorem by the general version (Theorem 1) allows a proof of the Lagrange Multiplier Theorem that weakens the classical C^1 hypothesis to just continuity of the constraint functions and their differentiability at the extremum [26]. As another example, comparative statics questions for equilibrium systems[20] can now be addressed with weaker assumptions.

4. Some historical background and comparisons

We originally developed the general theorems to solve an applied problem [26]. However, attempts to ascertain their novelty led to a historical study, and we present a few of the highlights that may be of interest to some readers.

We sketch the development of the implicit function theorem, with primary emphasis on the smoothness assumptions underlying the proofs. For the early period we have been guided by the information in Osgood [33], with Cauchy the earliest mentioned. We have not tried to examine the history prior to Cauchy's work. We use the notation of the previous sections.

There are several dimensions we could use for making historical comparisons. We could discuss assumptions: real scalars or complex scalars; the number of x and y variables; the smoothness assumptions on f with respect to x or y or (x, y); smoothness assumptions at $(\bar{x}, \bar{y})$ or in a neighborhood of the point. We could discuss conclusions: existence of implicit function, uniqueness of implicit functions, and smoothness of implicit functions. We could discuss methods of proof: Cauchy's calculus of residues, Cauchy's *Calcul des limites*, differential equations, or fixed point theorems.

We will give a brief chronological history, highlighting major differences among the various contributions, comparing our three theorems above with the major earlier results.

1831. Cauchy. Cauchy's results [10] are presumably the earliest rigorous existence proofs of the Implicit Function Theorem. Here the underlying scalar field K is that of the complex numbers. He mentions the general case in the introduction, and gives a detailed analysis of the special case $k = 1, n = 1$. The function f is assumed to be represented by a power series — i.e., to be analytic. A unique implicit function ϕ is obtained, and it is shown to be analytic.

Cauchy's tools were his Calculus of Residues and his Calculus of Limits. Their flavor may be experienced by considering the simple case where $k = 1$ and $n = 1$, so that both x and y are one-dimensional. Then a presentation of Cauchy's approach, in modern language looks roughly like the following.

---

[20] Cf. [37], Part I.

Assuming f is analytic and not identically zero, for any given x, the generalized Argument Principle implies:[21]

$$\frac{1}{2\pi i}\int_C y\frac{f_y(x,y)}{f(x,y)}dy = \sum_{r=1}^{p} \alpha_r a_r, \tag{91}$$

where the a_r are zeros of $f(x,\,\cdot\,)$ and the α_r are their multiplicities, and where C is a closed rectifiable curve not passing through any root a_r. Since the roots of an analytic function that is not identically zero are isolated, one can pick a small enough circle C in the complex plane about the root a_1 of $f(x,\,\cdot\,)$ so that there are no other zeros in the circle. In particular, Cauchy considers the case that the root a_1 has multiplicity 1.[22] Then denoting a_1 by $\phi(x)$, (91) becomes:[23]

$$\frac{1}{2\pi i}\int_C y\frac{f_y(x,y)}{f(x,y)}dy = \phi(x). \tag{92}$$

In view of the continuity of f and its derivatives, Rouché's Theorem[24] implies that for all nearby x the number of zero's (counting multiplicities) of the function $f(x,\,\cdot\,)$ remains the same; so there is a unique local solution function ϕ, and it can be calculated by the integral formula (92).

Cauchy further verifies that ϕ is (locally) an analytic function — that it is represented by its Maclaurin series. First he applies the calculus of residues to calculate derivatives for the Maclaurin series. Without loss of generality assuming $x = 0$ and $\phi(x) = 0$:[25]

$$\begin{aligned}
\phi^{(m)}(0) &= \frac{1}{2\pi i}\int_C y\left[\frac{\partial^m}{\partial x^m}\left(\frac{f_y(x,y)}{f(x,y)}\right)_{x=0}\right]dy \\
&= \frac{1}{2\pi i}\int_C yA_m(y)dy \quad \text{(defining } A_m) \\
&= \frac{1}{2\pi i}(2\pi i\,\mathrm{Res}(yA_m(y);0)) \quad \text{(by the Residue Theorem)} \\
&= \frac{1}{(m-1)!}\lim_{y\to 0}\frac{d^{m-1}}{dy^{m-1}}((y^m)yA_m(y)),
\end{aligned} \tag{93}$$

where the last equality follows[26] since $yA_m(y)$ has a pole of order m. Then he shows that the series converges to ϕ.

[21] Cf. [11], p. 124, Theorem 3.6.
[22] When the usual assumption is made that $f_y(\bar{x},\bar{y})$ is surjective, it follows that the unique root within the circle is a simple root.
[23] Cf. [10], p. 76 (52). Cauchy expresses y in polar coordinates, so his integrand has additional factor y.
[24] Cf. [11], p. 125, Theorem 3.8.
[25] Cf. [10], p. 83, equation (97).
[26] Cf. [11], p. 113, Proposition 2.4.

Of course this is an anachronistic account of his methods: the generalized Argument Principle, Rouché's Theorem, etc., had perhaps not yet crystallized as named theorems, and Cauchy established them in the context of his proof.[27]

1852. Cauchy. In his memoir [9], Cauchy used an alternative approach for proving the Implicit Function Theorem, namely existence theorems for differential equations. Again the underlying scalar field is that of the complex numbers, and the function f is analytic ("synectic"). The implicit function ϕ is unique and analytic. It is explicitly assumed that $f_y(\bar{x}, \bar{y})$ is nonsingular.

Although he mentions more general cases, formal proofs are given only for the case where x is one-dimensional but y can be multi-dimensional (i.e., $n = 1$ and $k \geqq 1$). In this situation Cauchy is able to apply existence theorems for ordinary differential equations to establish existence, uniqueness, and differentiability of the implicit function ϕ.

We can illustrate the logic of his approach again for the simple case $f(x, y) = 0$ with both x and y one-dimensional. Assume[28] that f is C^1 and that $f_y(\bar{x}, \bar{y}) \neq 0$. In that case, the right hand side of the ordinary differential equation

$$\frac{dy}{dx} = -\frac{f_x(x, y)}{f_y(x, y)} \tag{94}$$

with the initial condition $f(\bar{x}, \bar{y}) = 0$ is well defined and has a unique local solution, say $y = \phi(x)$, satisfying $\bar{y} = \phi(\bar{x})$. It can be shown that the function ϕ is the unique function locally satisfying $f(x, \phi(x)) = 0$ with $\bar{y} = \phi(\bar{x})$; i.e., it is the implicit function for the given function f. Furthermore, its derivative satisfies the relation

$$\phi'(x) = \frac{f_x(x, \phi(x))}{f_y(x, \phi(x))}. \tag{95}$$

Thus we have obtained the standard Implicit Function Theorem (for this simple case), but at the cost again of assuming the given function f to be C^1. Hence it could have been used to obtain Dini's result for a system involving a single independent variable. Perhaps an analogous approach using existence theorems for partial differential equations could be used when there are several independent variables, i.e., when $n \geqq 1$.

Theorem 1 above weakens the assumptions of both Cauchy papers by reducing the smoothness requirements on f (from analytic in a neighborhood to differentiable at a point). Although Theorem 1 obtains existence of an implicit function and asserts that all implicit functions are differentiable, it cannot claim uniqueness of the implicit function.

---

[27] A lucid exposition of the methods he used in [10] for establishing (92) is contained in [30].

[28] This logic applies both to the real scalars and also to the complex scalars (in which case C^1 is equivalent to analytic).

1875. Briot and Bouquet.[29] Briot and Bouquet in their [5] are clearly aware of and following Cauchy's work in the theory of complex variables. The scalars are complex, and f is analytic. In Section 211, p. 336, Theorem III, they state and prove the theorem for the special case $k = 1, n = 1$; in Section 212, p. 337, Theorem IV, they state and sketch an argument for the more general case $k \geqq 1$, but still with $n = 1$.

They use solution methods for ordinary differential equations to establish existence of an analytic implicit function. The uniqueness of the implicit function is implied by their results on ordinary differential equations (an explicit uniqueness argument is given in Section 209, p. 332, for the case of a single ordinary differential equation).

1877–78. Dini. In his lecture notes [15], Dini states the Implicit Function Theorem in the form found in most textbooks today. The scalars are real, and f is C^1 (or C^r for some $r \geqq 1$). He solves for a unique implicit function ϕ, which he proves is C^1 (respectively, C^r).

At that time, of course, Brouwer's Fixed Point Theorem [7] was not available.[30] Instead, Dini used the Intermediate Value Theorem together with a C^1 assumption. To indicate the essence of his approach, consider the simple case where both x and y are one-dimensional and there is only a single given function f and a single equation $f(x, y) = 0$.

In this context, Dini assumes that, in some neighborhood of $(\bar{x}, \bar{y})$, the function f is C^1 and furthermore, that the partial derivative $f_y(\bar{x}, \bar{y}) \neq 0$. His proof of the existence of an implicit function $y = \phi(x)$ then uses two propositions of differential calculus: the Intermediate Value Theorem and Taylor's Formula with Remainder (Extended Theorem of the Mean) in the form $f(\bar{x}+h, \bar{y}+k) = f(x, y) + hf_x(\bar{x}+\theta h, \bar{y}+\theta k) + kf_y(\bar{x}+\theta h, \bar{y}+\theta k)$, with $0 < \theta < 1$.

Thus even to establish existence, this proof uses the existence of the derivatives of f with respect to x and y in a neighborhood of $(\bar{x}, \bar{y})$. Furthermore it uses the continuity of the partial f_y to show that, in a sufficiently small neighborhood, it retains the same sign as at $(\bar{x}, \bar{y})$, and the continuity of f_x is also used.

The first printed version of Dini's implicit function theorem is in Genocchi's [17], as edited by Peano. (The German translation is [18].) A clear formulation of Dini's theorem and a proof is also found in [34], with a helpful diagram.

Theorem 1 weakens Dini's assumptions by allowing both real and complex scalars, and by weakening the smoothness requirements on f. Although Theo-

[29] (We have not had access to Briot and Bouquet's first edition of 1859. (Osgood [33] mistakenly dates the 2nd edition as 1873).)

[30] The paper itself is dated July, 1910.

rem 1 asserts that all the implicit functions are differentiable at $(\bar{x}, \bar{y})$, it cannot claim uniqueness of the implicit functions.

1893. Jordan. In Sections 91–95 of [27] the scalars are real, and the results do not seem to go beyond those of Dini. The general case $k \geqq 1, n \geqq 1$ is considered, f is C^1, and a unique implicit function ϕ is obtained. Jordan notes that its partial derivatives exist, and his discussion makes clear that they are continuous.

In Section 191, page 178, the scalars are complex. The case $k = 1, n \geqq 1$ is considered in detail; the possibility of an extension to $k \geqq 1$ is mentioned. The function f is analytic ("synectic"), and a unique implicit function ϕ is obtained, and it is shown to be analytic, with the usual formula for its partial derivatives.

A clear proof along the same lines, for $k = 1, n = 1$ is also to be found in [34], p. 345ff.

Our Theorem 1 weakens the analytic hypothesis on f to differentiability at $(\bar{x}, \bar{y})$, while retaining existence of an implicit function and differentiability of every implicit function at $(\bar{x}, \bar{y})$. It cannot, however, guarantee uniqueness or analyticity of the implicit function.

1899. Lindelöf. In [29] the scalars are complex. The general case $k \geqq 1, n \geqq 1$ is considered. The function f is assumed analytic, an implicit function ϕ is obtained, and it is shown to be analytic.

The proof is based on power series expansions of f and the implicit function ϕ, using Cauchy's *Calcul des limites*, in contrast to Jordan (1893), which reduced the complex case to the real case.

Our Theorem 1 weakens Lindelöf's assumption that f is analytic to differentiability at $(\bar{x}, \bar{y})$ and local continuity, while retaining existence of an implicit function. While it obtains differentiability of every implicit function at $(\bar{x}, \bar{y})$, it cannot claim uniqueness.

1901. Osgood. This contains a clear statement of an implicit function theorem for complex scalars, with $k \geqq 1, n \geqq 1$, and with an explicit statement of uniqueness of the implicit function. We have also found this to be a very useful source for historical information; see especially pp. 19 (footnote 30) and 103 (footnote 247).

1903. Goursat. In [19] the scalars are real, and $n \geqq 1$ and $k \geqq 1$. Goursat makes no differentiability assumption with respect to x. He assumes that f is continuous in a neighborhood of $(\bar{x}, \bar{y})$, and he retains the assumptions that $f(x, \cdot)$ is C^1 for x near 0. He obtains a unique implicit function, and shows that it is continuous.

Goursat's proof used a fixed point theorem, what we would now call a contraction mapping theorem, which he proved using Picard's method of successive approximations. Defining

$$g(x,y) = y - \bar{y} - (f_y(\bar{x},\bar{y}))^{-1} f(x,y) \qquad (96)$$

he shows that $g(x,y) = y$ has a unique solution for $y = \phi(x)$ that is a continuous function of x.[31]

Theorem 2 weakens Goursat's assumptions by allowing both real and complex scalars, and by weakening the smoothness requirements on f. Although Theorem 2 obtains existence of an implicit function and asserts that all implicit functions are continuous, it cannot claim uniqueness of the implicit functions.

The Goursat result (dropping differentiability of f with respect to x, and assuming continuity of f near $(\bar{x},\bar{y})$) is also found in Bliss's published lecture [3], pp. 8–9 on the "Fundamental Existence Theorems." Although Bliss cites Goursat's paper, his method of proof is based on Taylor's Theorem.

In [22] Graves notes that existence, uniqueness, and continuity can be obtained without assuming any differentiability of f with respect to the variable x. His remark (p. 139) is in the midst of his proof of his Theorem 2 (p. 138).

1909. W. H. Young. There are several theorems to consider in Young's [39]. In all of them, the scalars are the reals.

Young's Theorem 9 modifies Dini's result, replacing the C^r differentiability assumption on f with an assumption that f is differentiable of order r for some $r \geqq 2$. It asserts the existence of a unique implicit function ϕ, which is also differentiable of order r. Neither our Theorem 1 nor Dini's theorem require second order differentiability for existence of an implicit function.

Our Theorem 1 uses weaker assumptions than Young's Theorem 9 in allowing both real and complex scalars, and it uses weaker smoothness requirements on f (differentiability at $(\bar{x},\bar{y})$, instead of twice differentiable at $(\bar{x},\bar{y})$, hence C^1 in a neighborhood). Although our Theorem 1 obtains existence of an implicit function and asserts that all implicit functions are differentiable at $\bar{x}$, it cannot claim uniqueness of the implicit functions.

Young's Theorem 10 has a weaker assumption on f than Dini's theorem, since it requires a lower order differentiability of f with respect to y (differentiability in a neighborhood of $(\bar{x},\bar{y})$, rather than C^1). However it strengthens Dini's assumption that $f_y(\bar{x},\bar{y})$ is nonsingular (hence nonsingular in some neighborhood, since f is C^1 for Dini), by imposing nonsingularity conditions on certain principal minors (in some neighborhood). Young then concludes that there exists a unique implicit function ϕ, and it is differentiable at $(\bar{x},\bar{y})$.

Young remarks that only the uniqueness of ϕ is affected if one weakens the assumptions of his Theorem 10 by only requiring that the Jacobian and principal minor conditions hold at $(\bar{x},\bar{y})$, rather than throughout some neighborhood.[32]

[31] He notes that the solution is unique provided that g is Lipschitz continuous with constant $K < 1$, observing that this automatically holds when f_y is continuous in a neighborhood of $(\bar{x},\bar{y})$. (See also footnote 10, p. 78 above.)

[32] Page 420, paragraph 16.

Our Theorem 2 uses weaker assumptions than in Young's Theorem 10 in allowing both real and complex scalars, in only requiring continuity of f locally and differentiability at $(\bar{x}, \bar{y})$ (rather than differentiability locally), in not requiring Young's conditions on principal minors of the Jacobian in a neighborhood (it only requires that the Jacobian itself be nonsingular at $(\bar{x}, \bar{y})$). Our Theorem 2, however, cannot claim uniqueness of the implicit functions.

Young claims (in his Corollary 3 of Theorem 10) that Dini's result is a corollary of his Theorem 10, although that is not clear to us.

In neither Theorem 9 nor Theorem 10 does Young consider a situation in which f is not differentiable with respect to x, in contrast to Goursat's result and our Theorem 2. However, his Theorem 5, as extended in his section 10, contains a special case that does not require differentiability. For the special case $k = 1, n \geqq 1$, assuming only that $f_y(\bar{x}, \bar{y})$ is nonzero, and that f is continuous separately in each argument, he obtains in parts (1) and (2) the existence of an implicit function ϕ. (While there need not be a unique implicit function, he notes that there exists an upper semicontinuous one and a lower semicontinuous one.)

The basic idea of Young's proof is similar in spirit to what others have done. Because it yields a better result for the special case $k = 1, n \geqq 1$ than either his general Theorems 9 or 10, or our Theorem 2, we present here a proof sketch. (We weaken his continuity assumption with respect to x by only requiring continuity at $\bar{x}$.)

> Because $f(\bar{x}, \bar{y}) = 0$ and $f_y(\bar{x}, \bar{y}) \neq 0$, there exists a $\hat{y}$ such that $f(\bar{x}, \hat{y}) > 0$ and $f(\bar{x}, -\hat{y}) < 0$. Because $f(\,\cdot\,, \hat{y})$ is continuous at $\bar{x}$, then for all small enough x we have both $f(\bar{x} + x, \hat{y}) > 0$ and $f(\bar{x} + x, -\hat{y}) < 0$. Then continuity of $f(\bar{x} + x, \,\cdot\,)$ and the Intermediate Value Theorem imply $f(\bar{x} + x, y^*) = 0$ for some y^* in the interval $[-\hat{y}, \hat{y}]$. By continuity of $f(\bar{x} + x, \,\cdot\,)$, the set of such y^* is closed, so one way to define the value of an implicit function at x is to chose the maximum such y^*.

Other. Without aiming at completeness, we mention a few other treatments of the implicit function theorem. Careful statements of the classical (Dini) version are found in Bolza [4] and Carathéodory [8].

In the complex domain, Goursat [20], pp. 399 ff., [21], pp. 233 ff. and Osgood [35], §§45, 105, [35], pp. 86–86, and Markushevich [31] all utilize the Weierstrass Preparation Theorem [38], pp. 107–114.[33] This theorem deals with cases where the equation $f(x, y) = 0$ has roots $y(x)$ whose multiplicity m may equal or exceed 1, and is applicable to cases where $f_y(\bar{x}, \bar{y}) = 0$, and in this respect has broader applicability than the results established in the present paper. When $m = 1$, the Preparation Theorem can be used to prove the Implicit

[33] Published in 1886, but presented in lectures since 1860.

Function Theorem for the case where $f_y(\bar{x}, \bar{y}) \neq 0$ and f is analytic. This is shown most explicitly by Markushevich [31], p. 109, Theorem 3.11.

Evgrafov's proof [16] uses series expansions and Cauchy's integral formulae, as well as the Fréchet property of the differential. In addition to proving existence, uniqueness, and continuity, it also derives the formula $\phi'(x) = -f_y(x, y)f(x, y)$.

The paper by Hildebrand and Graves [24] was perhaps the first to use Banach space techniques to establish the Implicit Function Theorem through methods applying to both the real and complex domains.

Nijenhuis [32], building on a notion of Leach [28], obtained inverse and implicit functions in the real case, under an assumption ("strong differentiability") weaker than C^1, but stronger than Fréchet differentiability.

1974. Halkin. Halkin eliminated the C^1 hypotheses of earlier authors, using differentiability, continuity, and Brouwer's fixed point theorem [23], Theorem D. Unaware of his earlier result, we embarked on a similar quest, using very similar techniques. Though similar, some of our results differ from those of his theorems in a few significant respects.

We do not specify that the function f in Theorem 1 is continuous in a full neighborhood of $(\bar{x}, \bar{y})$; we only require continuity with respect to the y variable. In fact, we weaken even that continuity considerably, allowing continuity requirements to depend on the distance of the x variable from $\bar{x}$. And in Remark 1 we show that those continuity requirements are "tight." Halkin's Theorem D, however, assumes full continuity jointly in both variables, in a full neighborhood.

Halkin's Theorem D allows the space X to be a normed linear space, rather than the finite dimensional linear space we assume. (It appears that such a generalization of our result would be possible with only slight modifications of our proof.)

Our Theorem 2 is essentially the same as Halkin's Theorems A, B, and C, which do not require differentiability in the x variable.

Halkin does not have a parallel of our Theorem 3, which generalizes our Theorems 1 and 2.

Finally, our inverse function Theorem 4 is essentially the same as Halkin's Theorem G, and our proof obtains it from the implicit function theorem by the same standard techniques that he uses.

1989. Radulescu and Radulescu. Our inverse function Theorem 4 is essentially the same as Radulescu and Radulescu's Theorem 3.4 [36]. Their proof uses topological degree theory instead of the Brouwer fixed point theorem we use. While they do not state an implicit function theorem, that could be derived from their inverse function theorem by standard methods, but it would impose stronger continuity requirements than in our Theorem 1.

References

1. Berge, C.: Espaces Topologique. Dunod, Paris 1959
2. Berge, C.: Topological Spaces. Macmillan, New York 1963
3. Bliss, G.A.: Fundamental existence theorems. In: Lectures on Mathematics, pp. 1-107 American Mathematical Society 1913 (The Princeton Colloquium. Lectures delivered September 15 to 17, 1909)
4. Bolza, O.: Vorlesungen über Variationsrechnung. Chelsea Publishing Company, New York (second edition, no date, author's preface dated, 1909)
5. Briot, C.A.A., Bouquet, C.: Théorie des fonctions doublement périodiques et, en particulier des fonctions elliptiques. Gauthiers-Villars, Paris 1875 (Deuxième Édition)
6. Brouwer, L.E.J.: Beweis der Invarianz des n-dimensionalen Gebiets. Mathematische Annalen **71**, 305-313 (1912)
7. Brouwer, L.E.J.: Über Abbildung von Mannigfaldigkeiten. Mathematische Annalen **71**, 97-115 (1912)
8. Carathéodory, C.: Calculus of Variations and Partial Differential Equations of the First Order. Chelsea Publishing Company, New York, English edition, 1982 (Second (revised) edition originally published as Variationsrechnung und Partielle Differentialgleichungen erster Ordnung, B. G. Teubner, Berlin, 1935)
9. Cauchy, A.: Mémoir sur l'application du calcul infinitésimal à la détermination des fonctions implicites. In: Oeuvres Complètes d'Augustin Cauchy, volume 11, pp. 406-415. 1899. Edited by L'Académie des Sciences; C.R., T. 34, p. 265 (23 février 1852).
10. Cauchy, A.: Résumé d'un mémoir sur la mécanique cèleste et sur un nouveau calcul appelé calcul des limites. In:Oeuvres Complètes d'Augustin Cauchy, volume 12 of II, pp. 48-112. Gauthier-Villars, Paris, 1916. Edited by L'Académie des Sciences. The part pertaining to series expansions was previously read at a meeting of the Academie de Turin, 11 October, 1831.
11. Conway, J.B.: Functions of One Complex Variable. Springer-Verlag, New York, second edition, 1978.
12. Debreu, G.: A social equilibrium existence theorem. In: Proceedings of the National Academy of Sciences, volume 38, pp. 886-893 1952
13. Debreu, G.: Theory of Value. John Wiley and Sons, New York 1959
14. Dieudonné, J.: Foundations of Modern Analysis. Academic Press 1969 (Enlarged and Corrected Printing)
15. Dini, U.: Analisi Infinitesimale. Litografia Gorani, Pisa 1877-1878 (Lithographed version of manuscript lecture notes)
16. Evgrafov, M.A.: Analytic Functions. W.B. Saunders Company, Philadelphia 1966
17. Genocchi, A.: Calcolo Differenziale e Principii di Calcolo Integrale. Fratelli Bocca, Turin 1884 (Pubblicato con aggiunte dal Dr. Guiseppe Peano)
18. Genocchi, A.: Differentialrechnung und Grundzüge der Integralrechnung. Teubner, Leipzig 1899 (German translation by G. Bohlman and A. Schepp)
19. Goursat, É.: Sur la théorie des fonctions implicites. Bulletin de la Société Mathématique de France **31**, 184-192 (1903)
20. Goursat, É.: A Course in Mathematical Analysis, volume I. Ginn and Company, Boston 1904. Translated by E. R. Hedrick.
21. Goursat, É.: A Course in Mathematical Analysis, volume II, Part 1. Ginn and Company, Boston 1916. Translated by E. R. Hedrick and Otto Dunkel.

22. Graves, L.M.: The Theory of Functions of Real Variables. McGraw-Hill, New York 1956 (second edition)
23. Halkin, H.: Implicit functions and optimization problems without continuous differentiability of the data. Siam Journal on Control **12**, 229-236 (1974)
24. Hildebrand, H., Graves, L.M.: Implicit functions and their differentials in general analysis. Transactions of the American Mathematical Society **29**, 127-153 (1927)
25. Hurwicz, L.:, Richter, M.K.: Implicit Functions and Diffeomorphisms without C^1. Discussion Paper No. 279. Department of Economics, University of Minnesota 1994
26. Hurwicz, L., Richter, M.K:. Optimization and Lagrange Multipliers: Non-C^1 Constraints and "Minimal" Constraint Qualifications. Discussion Paper No. 280. Department of Economics, University of Minnesota 1995 (see also this volume, pp.97-151)
27. Jordan, C.: Cours d'analyse de l'École Polytechnique. (Tome Premier, Calcul différential, deuxième edition.) Gauthier-Villars et Fils, Paris eme edition 1893
28. Leach, N.B.: A note on inverse function theorems. Proceedings of the American Mathematical Society **12**, 694-697 (1961)
29. Lindelöf, E.: Demonstration élémentaire de l'existence des fonctions implicites. Bulletin des Sciences Mathématique **23**, 68-75 (1899) (Deuxième Série)
30. Lindelöf, E.: Le Calcul des Résidus et ses Applications à la Théorie des Fonctions. Chelsea Publishing Company, New York 1947 (Preface dated November 13, 1904)
31. Markushevich, A.I.: Theory of Functions of a Complex Variable, volume II. Prentice Hall, Englewood Cliffs, NJ 1965 (English edition) Revised translated and edited by R. A. Silverman.
32. Nijenhuis, A.: Strong derivatives and inverse mappings. American Mathematical Monthly **81**, 969-980 (1974)
33. Osgood, W.F.: Analytische Funktionen komplexer Grössen. In: Encyklopädie der Mathematischen Wissenschaften, volume Zweiter Teil. Analysis, pp. 1-114. Teubner, Leipzig 1901–1921 (Originally published in "Heft 1," whose date was 27.XII.1901)
34. Osgood, W.F: Lehrbuch der Funktionentheorie, volume I. Teubner, Leipzig 1907
35. Osgood, W.F.: Lehrbuch der Funktionentheorie, volume II.1. Teubner, Leipzig 1929 (second edition)
36. Radulescu, S., Radulescu, M.: Local inversion theorems without assuming continuous differentiability. Journal of Mathematical Analysis and Applications **138**, 581-590 (1989)
37. Samuelson, P.A.: Foundations of Economic Analysis. Harvard University Press 1947
38. Weierstrass, K.: Abhandlungen aus der Functionenlehre. Julius Springer, Berlin 1886
39. Young, W.H.: On implicit functions and their differentials. In: Proceedings of the London Mathematical Society, volume 7, pp. 397-421 1909 (Second Series)

Adv. Math. Econ. 5, 97–151 (2003)

Advances in
MATHEMATICAL
ECONOMICS

Optimization and Lagrange multipliers: non-C^1 constraints and "minimal" constraint qualifications *

Leonid Hurwicz[1] **and Marcel K. Richter**[2]

[1] Department of Economics, University of Minnesota, 271 19th Avenue South, Minneapolis, MN 55455, USA
(e-mail: hurwicz@tc.umn.edu)
[2] Department of Economics, University of Minnesota, 271 19th Avenue South, Minneapolis, MN 55455, USA
(e-mail: richter@econ.umn.edu)

Received: August 1, 2002
Revised: August 22, 2002

JEL classification: C600, C650

Mathematics Subject Classification (2000): 26B10, 57R50

Abstract. Constrained optimization problems are central to economics, and Lagrange multipliers — when they exist — play a basic role in solving them, in theory and in practice. Examples are well known of optimization problems for which multipliers do not exist. So it is important to know what requirements constraint functions must satisfy to be "Lagrange regular," i.e. to guarantee existence of multipliers for broad classes of maximand or minimand functions. We relax the requirements in three directions:

1. We reduce the smoothness requirements on constraints. This allows weaker and more uniform hypotheses for mixed inequality and equality constraints, permitting, for example, just differentiability at the optimum and continuity in a neighborhood. (We allow much weaker hypotheses, as well.)

   Beyond smoothness, other requirements have long been imposed on constraint functions, to avoid simple examples lacking multipliers. We examine two types of such "constraint qualifications".
2. We provide new, relaxed constraint qualifications of both Jacobian and path types.
   a) Our Jacobian constraint qualifications (23), (24), (25) permit spanning properties as alternatives to the usual rank restrictions.

* For valuable comments on an earlier version, we are indebted to Professor Kam-Chau Wong, Chinese University of Hong Kong; Professor Hülya Eraslan, University of Pennsylvania; and Nevzat Eren, University of Minnesota.

   b) Our path constraint qualifications (69), (72), (73) impose fewer restrictions than before on the directions permitted for constraint derivatives. (The logical relationships are indicated in (149)).
3. Our relaxed requirements are not only sufficient for avoiding many well-known counterexamples — they cannot be weakened further:
   c) We formalize a notion of minimality for Jacobian constraint qualifications, and prove that ours are minimal for "Lagrange regularity."
   d) We prove that our path constraint qualifications are necessary for "Lagrange regularity."

The tool enabling us to relax smoothness requirements on equality constraints is our Non-C^1 Implicit Function Theorem, p. 142. (See also the reference in Section 8. to Halkin's work.)

Since we are weaving together many strands, the following outline may be helpful.

1. Introduction: Constraint qualifications, definitions	page 98
2. Constrained maximization	page 100
Lagrange regularity	page 101
Fundamental Lemma	page 104
3. The Jacobian criterion	page 106
4. The Jacobian criterion is sufficient	page 110
Theorem 1	page 110
Theorem 1^+	page 114
5. The Jacobian criterion is minimal	page 120
Theorem 2	page 120
6. The Tangency-path criterion	page 124
Theorem 3 (Sufficiency)	page 127
Theorem 4 (Necessity)	page 132
Weaker than Jacobian criterion	page 132
7. Mathematical Appendix	page 138
8. Historical comments and comparisons	page 143

Key words: Constrained optimization, Lagrange, Kuhn-Tucker, non-C^1 analysis, minimal constraint qualification, Jacobian Criterion, Tangency-Path Criterion

1. Introduction

Lagrange-Kuhn-Tucker multipliers are important theoretical and practical tools for solving many types of constrained optimization problems. It is well known, however, that such multipliers do not exist for all problems — that certain requirements have to be imposed on the constraints, in order to be assured that every maximand or minimand function will generate multipliers. In this paper we seek to reduce those requirements as much as possible in two directions: smoothness requirements, and "constraint qualifications."

(1) Smoothness of constraints. While the classical Lagrange Multiplier Theorem assumes C^1 smoothness of equality constraints and a (rank) *Jacobian-type condition*, Kuhn and Tucker reduced that to differentiability for inequality constraints, using a *path-type condition.* Is a similar reduction possible for equality constraints or mixed constraints? As shown below, the answer is yes.[1,2] For Jacobian-type constraint qualifications we show that it can be reduced to differentiability at just the optimizing point, together with partial continuity. For path-type constraint qualifications we reduce the smoothness to just existence of partial derivatives.

(2) Constraint qualifications. For optimality problems with inequality constraints, it has been known since Karush that some "constraint qualifications" were required to guarantee existence of "Lagrange multipliers." In the Karush-Kuhn-Tucker inequality theorems, these were properties of paths lying in the constraint set. For problems with equality constraints, corresponding qualifications were (rank) properties of the constraint Jacobian, implicit already in Lagrange. For both Jacobian and path conditions, our results are achieved under weaker assumptions than found elsewhere. We also introduce a notion of minimality: our Jacobian condition is minimally sufficient for existence of Lagrange multipliers; finally, we prove our path condition is necessary and sufficient.

Type 1 results: Theorems 1, 1^+, 3. (Weak smoothness hypotheses and weak constraint qualifications sufficient for Lagrange regularity.) Suppose that $\bar{u} \in I\!R^n$ maximizes the real valued function f subject to $g \geqq 0$ and $h = 0$. Then under either Condition (a) or Condition (b) below, there exist Lagrange multipliers $\lambda \geqq 0$ in $I\!R^p$ and μ in $I\!R^k$ such that:

$$f'(\bar{u})^T + \lambda^T g'(\bar{u}) + \mu^T h'(\bar{u}) = 0, \tag{1a}$$

i.e., such that:

$$\begin{aligned} &\frac{\partial f}{\partial u_i}(\bar{u}) + \lambda_1 \frac{\partial g^1}{\partial u_i}(\bar{u}) + \ldots + \lambda_p \frac{\partial g^p}{\partial u_i}(\bar{u}) \\ &\qquad + \mu_1 \frac{\partial h^1}{\partial u_i}(\bar{u}) + \ldots + \mu_k \frac{\partial h^k}{\partial u_i}(\bar{u}) = 0 \qquad (i = 1, \ldots, n). \end{aligned} \tag{1b}$$

Condition a) f, g and h are differentiable at $\bar{u}$, and h in addition satisfies certain partial continuity conditions there, and the Jacobian $(g'(\bar{u}), h'(\bar{u}))$ has certain properties.

[1] Halkin [16] has an earlier result in this direction. Although there are many points of similarity between his approach and parts of ours, his results concern only "quasi-regularity" (p. 103 below), rather than the full regularity of our Theorems 1, 1^+; that much weaker and less informative conclusion enables him to avoid all mention of constraint qualifications, which are the main focus of all our theorems. Also, his smoothness assumptions are more restrictive than those of our Theorem 1^+(A,B). See also footnote 29 and the Historical Notes and Comparisons in Section 8..

[2] We restrict attention to constraints and optimizing functions that have certain types of derivatives at the optimizing point. We do not consider subdifferentials, which would become more natural if we restricted attention to constraints that were concave or convex functions.

Condition b) f, g, and h admit partial derivatives, and satisfy a certain "path-type" constraint qualification.

Note that the "constraint qualifications" (a) and (b) refer only to the constraint functions g and h, not to the maximand f. When multipliers exist for g and h for a whole class $\mathcal{F}$ of functions f, we say that (g, h) is "Lagrange regular" for $\mathcal{F}$.

Type 2 results: Theorems 2, 4. (Minimality of constraint qualifications.) To guarantee Lagrange regularity of (g, h) for a class $\mathcal{F}$ of maximands:

i) The properties listed in Condition (a) are necessary for Lagrange regularity, if we restrict ourselves to conditions expressed in terms of the Jacobians of the constraint functions, i.e. in terms of partial derivatives.
ii) The properties listed in Condition (b) are necessary for Lagrange regularity.

Our relaxations of constraint qualifications listed in Conditions (a) and (b) are precisely those needed to make the constraint qualifications minimally sufficient in (a), and necessary and sufficient in (b), respectively. In other words, if Condition (a) were violated, then the existence of Lagrange multipliers could not be decided by looking at only the Jacobians of g and h. And if Condition (b) were violated, then the existence of Lagrange multipliers could not be decided by looking at only the constraints g and h: one would have to look at the particular maximand f as well.

2. Constrained maximization

For simplicity, we study just constrained maximization, since the minimization of a function f can be studied in terms of maximizing $-f$. And we discuss nonnegativity constraints $g \geqq 0$, since nonpositivity constraints can be studied through $-g \geqq 0$.

We seek to maximize a real valued function f on a set $U \subseteqq I\!R^n$, subject to conditions $g^1 \geqq 0, \ldots, g^m \geqq 0$ and $h^1 = 0, \ldots, h^k = 0$, where the g^i and h^i are also functions from U to $I\!R^n$. Either g or h, or both, may be absent. To avoid trivialities we assume throughout that $n > 0$.

Let $g = (g^1, \ldots, g^m)$ and $h = (h^1, \ldots, h^k)$, and define the *constraint set* by

$$C(g,h) = \{x \in U \colon \forall i_{\,i=1,\ldots,m}\; g^i(x) \geqq 0 \;\;\&\;\; \forall j_{\,j=1,\ldots,k}\; h^j(x) = 0\}. \quad (2)$$

A point $\bar{u} \in U$ is said to *maximize* f *on* $C(g, h)$ if:

$$\begin{aligned} &\bar{u} \in C(g,h) \\ &\forall u_{\,u \in C(g,h)}\; f(\bar{u}) \geqq f(u). \end{aligned} \quad (3)$$

When g is absent, this is a problem of classical mathematics; when h is absent, it is known in economics as a "Kuhn-Tucker" problem. We write $C(g)$ when h is absent, and $C(h)$ when g is absent; the context will avoid ambiguity. When both g and h are absent (so there are no constraints), then $C(g,h) = U$.

2.1 Lagrange regularity

Suppose that $\bar{u} \in U$ maximizes f on $C(g,h)$ and that f, g, and h have partial derivatives at $\bar{u}$. We are interested in the existence of "Lagrange multipliers," $\lambda \in I\!R^m$ and $\mu \in I\!R^k$, satisfying:

$$f'(\bar{u})^T + \lambda^T g'(\bar{u}) + \mu^T h'(\bar{u}) = 0. \tag{4}$$

In addition, we will want λ to satisfy a nonnegativity condition. But simple examples show that there may exist no λ and μ satisfying (4). Consider:[3]

$$\begin{aligned} g^1(u_1,u_2) &= u_2 - u_1^2 \geqq 0 \\ g^2(u_1,u_2) &= -u_2 - u_1^2 \geqq 0, \end{aligned} \tag{5}$$

or:

$$\begin{aligned} h^1(u_1,u_2) &= u_2 - u_1^2 = 0 \\ h^2(u_1,u_2) &= -u_2 - u_1^2 = 0. \end{aligned} \tag{6}$$

In each case the constraint set is $\{0\}$, so any function $f(u_1,u_2)$ has a constrained maximum at $\bar{u} = 0$. But $f(u_1,u_2) = u_1$ admits no λ or μ satisfying (4).

So special assumptions are needed to guarantee existence of multipliers λ and μ. To state them, we first explain the notion of a "binding" inequality constraint.

Binding constraints. An inequality constraint g^i is *binding at* $\bar{u}$ if $g^j(\bar{u}) \geqq 0$ for all $j = 1,\ldots,m$, and if $\bar{u}$ belongs to the boundary of the "individual" constraint set $\{x \in I\!R^n : g^i(x) \geqq 0\}$.[4]

With $U \subseteqq I\!R^n$, with $\bar{u} \in U$ and $g\colon U \to I\!R^m$, we partition $M = \{1,\ldots,m\}$ to distinguish the g^i that are binding at $\bar{u}$ from the others:[5]

---

[3] This is analogous to Slater's example in [38].

[4] The binding property refers to the boundary of the domain of g^i, not of its range. Thus g^i need not be binding at $\bar{u}$ even if $g^i(\bar{u}) = 0$ (e.g. if $g^i = 0$ in an open neighborhood of $g^i(\bar{u})$), and conversely, g^i can be binding at $\bar{u}$ even if $g^i(\bar{u}) > 0$ (e.g. if $U = \{(u_1,u_2) \colon u_1 + u_2 \geqq 2\}$, and $g^i(u_1,u_2) = u_1 + u_2$, and $(\bar{u}_1,\bar{u}_2) = (1,1)$).

Note that if $g^1 = g^2$, then *each* constraint may be binding at $\bar{u}$.

[5] One could introduce an analogous distinction between binding and nonbinding equality constraints, but we shall not do so.

$$I = \{i \in M\colon \bar{u} \text{ is in the boundary of } \{u \in U\colon g(\bar{u}) \geqq 0 \;\&\; g^i(u) \geqq 0\}\}$$
$$J = M \setminus I \tag{7}$$

(either I, J, or both may be empty). When I or J is nonempty, we write, respectively:

$$\begin{aligned} g_I &\colon U \to I\!R^p \\ g_J &\colon U \to I\!R^{m-p} \\ g = (g_I, g_J) &\colon U \to I\!R^p \times I\!R^{m-p} = I\!R^m. \end{aligned} \tag{8}$$

Without loss of generality, we assume that the components of g_I are the first p elements of g: $g_I = (g^1, \ldots, g^p)$.

Regularity. Now we seek conditions under which the fact that $\bar{u} \in U \subseteqq I\!R^n$ maximizes f on $C(g,h)$ implies the existence of a $\lambda \in I\!R^m$ and a $\mu \in I\!R^k$ such that:[6,7]

$$f'(\bar{u})^T + \lambda^T g'(\bar{u}) + \mu^T h'(\bar{u}) = 0, \tag{9}$$

$$\lambda \geqq 0 \tag{10a}$$

and

$$\lambda_i = 0 \qquad \text{if } g^i \text{ is not binding at } \bar{u} \quad (\text{i.e., } i \in J). \tag{10b}$$

Rather than solving (9) and (10) for λ and μ, given particular g^i, h^i, and f, one may ask whether the given functions g^i and h^j allow us to solve, for *each* f in some family $\mathcal{F}$ that is maximized on $C(g,h)$ at $\bar{u}$, for such λ and μ. (We shall always assume that each $f \in \mathcal{F}$ has partial derivatives at $\bar{u}$.) So we say that (g,h) is *Lagrange mixed-regular for* $(\bar{u}, \mathcal{F})$ if:[8]

i) all the g^i and h^j all have partial derivatives at $\bar{u}$;

---

[6] It is tempting to suppose that (10) implies $\lambda_i > 0$ whenever g^i is binding at $\bar{u}$. Simple examples show that is not true. Consider this example in $I\!R^1$: $f(x) = -x^2$, $g(x) = x$, and $\bar{u} = 0$. The constraint g is binding at the maximizer $\bar{u} = 0$, but $\lambda = 0$ is required by (9).

[7] When U is open (as we assume in Parts 4 and 5), then if g^i is binding at $\bar{u}$ it follows that $g^i(\bar{u}) = 0$. So for open U, condition (10) implies (though is not implied by) this "complementary slackness" condition:

$$\lambda_i \geqq 0 \qquad \lambda_i g^i(\bar{u}) = 0.$$

[8] The terminology is modified from [4]. We really should call this *Lagrange regularity for* $(U, \bar{u}, \mathcal{F})$; for brevity, we omit the reference to U. (Alternatively, we can note that U is determined by $\mathcal{F}$.)

ii) for every function $f \in \mathcal{F}$, if $\bar{u}$ maximizes f on $C(g, h)$ then (9) and (10) hold for some (λ, μ).

When h is absent, we say that g is *Lagrange inequality-regular* for $(\bar{u}, \mathcal{F})$ if:

i) all the g^i have partial derivatives at $\bar{u}$;
ii) for every function $f \in \mathcal{F}$, if $\bar{u}$ maximizes f on $C(g)$ then:

$$f'(\bar{u})^T + \lambda^T g'(\bar{u}) = 0 \tag{11}$$

and (10) hold for some λ.

And when g is absent we say that h is *Lagrange equality-regular* for $(\bar{u}, \mathcal{F})$ if:

i) all the h^j have partial derivatives at $\bar{u}$;
ii) for every function $f \in \mathcal{F}$, if $\bar{u}$ maximizes f on $C(h)$ then:

$$f'(\bar{u})^T + \mu^T h'(\bar{u}) = 0 \tag{12}$$

for some μ.

When for each $f \in \mathcal{F}$ there are unique λ and μ satisfying (9) and (10), we say that (g, h) is *Lagrange mixed-regular for* $(\bar{u}, \mathcal{F})$, *uniquely*; and we use similar uniqueness definitions for the inequality and the equality cases.

We often refer simply to Lagrange regularity; the context will make the type clear. When both g and h are absent, we say that we have *Lagrange regularity for* $(\bar{u}, \mathcal{F})$ if: for all $f \in \mathcal{F}$, if $\bar{u}$ maximizes f on U then $f'(\bar{u}) = 0$.

It is important to note that Lagrange regularity is a property of particular constraint functions g and h — not just a property of the constraint set $C(g, h)$. We demonstrate this in Remark 6.1.iii, page 126 below, and we note some implications in the Removable Constraints section, page 109 below.

Quasi-regularity. A weaker notion of regularity, apparently dates back to Hilbert (for equalities). We say that (g, h) is *Lagrange quasi mixed-regular for* $(\bar{u}, \mathcal{F}(\bar{u}))$ if the properties of Lagrange regularity hold except that conditions (9) and (10) are replaced by the weaker condition: there exists a nonzero vector $(\lambda_0, \lambda, \mu)$ such that

$$\lambda_0 f'(\bar{u})^T + \lambda^T g'(\bar{u}) + \mu^T h'(\bar{u}) = 0, \tag{13}$$

$$\lambda_0 \geqq 0 \tag{14}$$

$$\lambda \geqq 0. \tag{15}$$

We assumed that partial derivatives of the g^i, h^j, and f exist at $\bar{u}$, even when not binding, in order to give meaning to (9). We could avoid assuming that the nonbinding g^i possess partial derivatives at $\bar{u}$, by using (16) and (17) below in place of (9) and (10). This would be more general, but it would be less convenient to apply where the structure of the constraint set is not known a priori.

Similar definitions apply for inequalities and for equalities.[9]

In this paper we focus on regularity, as it is more informative than quasi-regularity. However, Lagrange quasi mixed-regularity implies Lagrange mixed-regularity, under the special rank assumption (31) below. This follows from the Fundamental Lemma below, on purely algebraic grounds.

Now we seek λ and μ that satisfy (9) and (10) for *all* such f. We begin by restating the problem in a simpler form with a natural

Reduction. Let $g: U \to I\!R^m$, $h: U \to I\!R^k$, and $f: U \to I\!R$ have partial derivatives at $\bar{u}$.[10] Suppose that the binding constraints are $g_I = (g^1, \dots, g^p)$. Then it is clear that a solution $\lambda \in I\!R^m$ and $\mu \in I\!R^k$ to (9) and (10) exists if and only if there exists a $\tilde{\lambda} \in I\!R^p$ and a $\mu \in I\!R^k$ such that:

$$f'(\bar{u})^T + \tilde{\lambda}^T g_I'(\bar{u}) + \mu^T h'(\bar{u}) = 0, \tag{16}$$

and such that:

$$\tilde{\lambda} \geqq 0. \tag{17}$$

2.2 The fundamental lemma. Constraint qualifications

Although our basic maximization question is one of analysis, it is useful to view it algebraically. Our answers all rest on the following simple consequence of the Theorem of the Alternative.

Lemma 2.1 (Fundamental lemma). *Let $U \subseteqq I\!R^n$, and let $\bar{u} \in U$ be a limit point.*

A) Suppose $f: U \to I\!R$, $g: U \to I\!R^p$, and $h: U \to I\!R^k$ are functions on U that have partial derivatives at $\bar{u}$, with $g(\bar{u}) = 0$ and $h(\bar{u}) = 0$, and suppose there does not *exist $(\lambda, \mu) \in I\!R^p \times I\!R^k$ satisfying the following two conditions:*

i)

$$f'(\bar{u})^T + \lambda^T g'(\bar{u}) + \mu^T h'(\bar{u}) = 0, \tag{18}$$

ii)

$$\lambda \geqq 0. \tag{19}$$

---

[9] Halkin [16] used an approach similar to those of [9], [5], [6], and [24], using Jacobian conditions to establish Lagrange quasi mixed-regularity under weaker smoothness hypotheses than those predecessors.

[10] Actually, the $g^{p+1}, \dots, g^m$ need not have partial derivatives, since they don't appear in (16), and they effectively don't appear in (9) since the corresponding λ_i vanish according to (10b).

Then there exists a $\tilde{z} \in I\!R^n$ *such that:*

$$g'(\bar{u})\tilde{z} \geqq 0 \tag{20a}$$

$$h'(\bar{u})\tilde{z} = 0 \tag{20b}$$

$$f'(\bar{u})\tilde{z} > 0. \tag{20c}$$

B) When g is absent, part (A) holds if we delete references to, and terms and relations involving λ *or g.*

C) When h is absent, part (A) holds if we delete references to, and terms and relations involving μ *or h.*

Proof. This is an almost immediate application of the Theorem of the Alternative.[11] □

The lemma is an algebraic statement about derivatives. It does not mention maximization, but now we show why it is basic to maximization theory.

Suppose that f, g, and h are differentiable at $\bar{u}$, which maximizes f on the constraint set $C(g,h)$. If (g,h) is not Lagrange regular, then the lemma implies that (20) holds. If there were a path $u(\cdot)$ with values $u(t)$ lying in $C(g,h)$, starting at $\bar{u}$ and with derivative $u'(0) = \tilde{z}$, then[12] we could rewrite (20) as:

$$\frac{d}{dt}g(u(t))|_{t=0} = g'(\bar{u})\tilde{z} \geqq 0 \qquad \text{(by (20a))} \tag{21a}$$

$$\frac{d}{dt}h(u(t))|_{t=0} = h'(\bar{u})\tilde{z} = 0 \qquad \text{(by (20b))} \tag{21b}$$

$$\frac{d}{dt}f(u(t))|_{t=0} = f'(\bar{u})\tilde{z} > 0 \qquad \text{(by (20c))}. \tag{21c}$$

So small movements along the path would keep us in the constraint set but increase the value of f. And that would contradict the assumption that $\bar{u}$ maximizes f on the constraint set $C(g,h)$.

Thus algebra highlights the role of analysis. As (5) and (6) show, Lagrange regularity requires some restrictions on the constraint functions g and h. As the algebra suggests, a sufficient restriction is the existence of paths $u(\cdot)$ with $u'(0) = \tilde{z}$ for any $\tilde{z}$ satisfying (20). For historical reasons, such hypotheses are typically called *constraint qualifications.*[13]

Two general types of constraint qualifications have been developed.[14] The first simply asserts that any $\tilde{z}$ satisfying (20a,b) is indeed the derivative of some path lying in $C(g,h)$. We will call these *path conditions*.

---

[11] See the Appendix.

[12] If g, h, and f were differentiable at $\bar{u}$.

[13] Cf. [30].

[14] Slater [38] proposed a constraint qualification that does not fall into the two types we discuss. It was intended, however, for application to the Saddle Point Equivalence Theorem, assuming concavity of the constraint functions g^i, rather than for a

The second type is computationally oriented, being expressed as algebraic properties of the Jacobian matrix $(g'(\bar{u}), h'(\bar{u}))$. They may involve its nonsingularity, the span of its rows, or other algebraic aspects. We call these *Jacobian conditions*.

In simple examples (e.g. (5) and (6)) checking whether a path condition holds can be rather easy. In more complicated examples, it may be difficult. By contrast, the Jacobian conditions that we discuss are decidable in an algorithmic fashion, as we will see later.

3. The Jacobian criterion

Our Jacobian condition will guarantee that every $\tilde{z}$ satisfying (20a,b) (or the corresponding variant when g or h is absent) is the derivative of some path lying in $C(g,h)$ (or $C(g)$ or $C(h)$). In Theorem 1 we prove that the condition is sufficient for Lagrange regularity. Then in Theorem 2 we prove it is as weak as possible in the class of sufficient Jacobian conditions. (It is generally not as weak, however, as the non-Jacobian path conditions in Theorems 3 and 4, which are both necessary and sufficient for Lagrange regularity.) In order to make these notions precise, we use the following terminology.

Sufficient and minimal Jacobian constraint qualifications. Let $U \subseteqq \mathbb{R}^n$, and let $\bar{u} \in U$ be a limit point. Let $\mathcal{G}$ be a set of functions $g\colon U \to \mathbb{R}^m$, let $\mathcal{H}$ be a set of functions $h\colon U \to \mathbb{R}^k$, and let $\mathcal{F}$ be a set of functions $f\colon U \to \mathbb{R}$. Assume that all functions in $\mathcal{G}$, $\mathcal{H}$, and $\mathcal{F}$ have partial derivatives at $\bar{u}$.

By a *Jacobian mixed-constraint qualification for* $(\bar{u}, \mathcal{G}, \mathcal{H})$ we mean a property $\mathbb{Q}$ of some members of the set $\{\begin{pmatrix} g'_I(\bar{u}) \\ h'(\bar{u}) \end{pmatrix} : g \in \mathcal{G} \;\; \& \;\; h \in \mathcal{H}\}$ of $(p+k) \times n$ (real) matrices. To each property $\mathbb{Q}$ corresponds the set Q of matrices with that property.

We say that a Jacobian mixed-constraint qualification $\mathbb{Q}$ for $(\bar{u}, \mathcal{G}, \mathcal{H})$ is $(\bar{u}, \mathcal{G}, \mathcal{H}, \mathcal{F})$*-sufficient for Lagrange mixed-regularity* if: for all $g \in \mathcal{G}$ and $h \in \mathcal{H}$,

$$\begin{pmatrix} g'_I(\bar{u}) \\ h'(\bar{u}) \end{pmatrix} \in Q \;\Rightarrow\; (g,h) \text{ is Lagrange mixed-regular for } (\bar{u}, \mathcal{F}). \tag{22}$$

We say that a Jacobian constraint qualification $\mathbb{Q}$ for $(\bar{u}, \mathcal{G}, \mathcal{H})$ is *minimally* $(\bar{u}, \mathcal{G}, \mathcal{H}, \mathcal{F})$*-sufficient for Lagrange mixed-regularity* if:

---

proof of Lagrange regularity. Because of the convexity of the constraint set defined by such g^i, Slater's condition implies a path-type constraint qualification of the type we discuss below. (See the historical comments in Section 8. below.)

i) $\mathbb{Q}$ is $(\bar{u}, \mathcal{G}, \mathcal{H}, \mathcal{F})$-sufficient for Lagrange mixed-regularity in the sense of (22);

ii) no weaker Jacobian property (i.e., no proper superset of Q) is $(\bar{u}, \mathcal{G}, \mathcal{H}, \mathcal{F})$-sufficient for Lagrange mixed-regularity: if $g \in \mathcal{G}$, $h \in \mathcal{H}$, and $\begin{pmatrix} g_I'(\bar{u}) \\ h'(\bar{u}) \end{pmatrix} \notin Q$, then there is some $\hat{g} \in \mathcal{G}$ and some $\hat{h} \in \mathcal{H}$ with $\hat{g}_I'(\bar{u}) = g_I'(\bar{u})$ and $\hat{h}'(\bar{u}) = h'(\bar{u})$ for which $(\hat{g}, \hat{h})$ is not Lagrange mixed-regular for $(\bar{u}, \mathcal{F})$.

We may abbreviate "minimally sufficient" to "minimal."

Analogous definitions of sufficiency and minimality apply with functions g for the inequalities problem, and with functions h for the equalities problem.

Whether a Jacobian condition is minimally sufficient for Lagrange regularity clearly depends on the set U, the element $\bar{u}$, and the classes $\mathcal{G}$ and $\mathcal{H}$ of equality and inequality constraint functions under consideration, as well as on the class $\mathcal{F}$ of maximand functions. To simplify discussing these classes, some notation is useful.

In the following list, U is a subset of $\mathbb{R}^n$. For any $\bar{u} \in U$ we make these definitions:

$\mathcal{F}_D(\bar{u})$ is the set of functions $f\colon U \to \mathbb{R}$ that are differentiable at $\bar{u}$.

$\mathcal{G}_D(\bar{u})$ is the set of functions $g\colon U \to \mathbb{R}^m$ that are differentiable at $\bar{u}$.

$\mathcal{H}_P(\bar{u})$ is the set of functions $h\colon U \to \mathbb{R}^k$ that have partial derivatives at $\bar{u}$.

$\mathcal{H}_D(\bar{u})$ is the set of functions $h\colon U \to \mathbb{R}^k$ that are differentiable at $\bar{u}$.

$\mathcal{H}_{DC}(\bar{u})$ is the set of functions $h\colon U \to \mathbb{R}^k$ that are differentiable at $\bar{u}$ and continuous in some neighborhood of $\bar{u}$.

$\mathcal{H}_{C^1}(\bar{u})$ is the set of functions $h\colon U \to \mathbb{R}^k$ that are C^1 at $\bar{u}$.

For example, to say that a Jacobian equality-constraint qualification $\mathbb{Q}$ is $(\bar{u}, \mathcal{H}_{DC}(\bar{u}), \mathcal{F}_D(\bar{u}))$-sufficient means: for all $h \in \mathcal{H}_{DC}(\bar{u})$, if $h'(\bar{u}) \in \mathbb{Q}$ then h is Lagrange equality-regular for $(\bar{u}, \mathcal{F}_D(\bar{u}))$.

The rest of this section states a new Jacobian constraint qualification, the Jacobian Criterion. Section 4. will prove it is sufficient for Lagrange regularity with respect to some important classes $\mathcal{G}$, $\mathcal{H}$, and $\mathcal{F}$. And Section 5. will prove it is minimally sufficient with respect to those same classes (and to other classes as well).

The Jacobian criterion. Suppose that $\bar{u} \in U \subseteqq \mathbb{R}^n$ and that[15] $g_I : U \to \mathbb{R}^p$ and $h\colon U \to \mathbb{R}^k$ have partial derivatives at $\bar{u}$. Let $a(1), \ldots, a(p)$ be the rows of the $p \times n$ Jacobian matrix $g_I'(\bar{u})$, and let $b(1), \ldots, b(k)$ be the rows of

[15] I is defined in (7).

the $k \times n$ Jacobian matrix $h'(\bar{u})$. We allow $p = 0$ or $k = 0$, or both; i.e. g_I or h may be absent.

A) (Mixed: Equalities and inequalities) We say (g, h) satisfies the *(Mixed-Problem) Jacobian Criterion at* $\bar{u}$ if one of the following two mutually exclusive conditions (a), (b) holds:

a) $\text{rank}(h'(\bar{u})) = k < n,$ (23a)
and there exists a $\xi \in I\!R^n$ such that:

$$g'_I(\bar{u})\xi > 0$$
$$h'(\bar{u})\xi = 0;$$

b) $\text{wedge}(a(1), \ldots, a(p)) + \text{span}(b(1), \ldots, b(k)) = I\!R^n,$ (23b)

i.e.:

$$\{v \in I\!R^n : \exists t_{\ 0 \leqq t \in I\!R^p} \ \exists z_{\ z \in I\!R^k} \ v = t_1 a(1) + \cdots + t_p a(p) + z_1 b(1) + \cdots + z_k b(k)\} = I\!R^n.$$

B) (Inequalities) We say g satisfies the *(Inequality-Problem) Jacobian Criterion at* $\bar{u}$ if one of the following two mutually exclusive conditions holds:

a) I is nonempty and there exists a $\xi \in I\!R^n$ such that: (24a)

$$g'_I(\bar{u})\xi > 0;$$

b) $\text{wedge}(a(1), \ldots, a(p)) = I\!R^n$, i.e.: (24b)

$$\{v \in I\!R^n : \exists t_{\ 0 \leqq t \in I\!R^p} \ v = t_1 a(1) + \cdots + t_p a(p)\} = I\!R^n.$$

C) (Equalities) We say that h satisfies the *(Equality-Problem) Jacobian Criterion at* $\bar{u}$ if one of the following two mutually exclusive conditions holds:

a) $\text{rank}(h'(\bar{u})) = k < n;$ (25a)
b) $\text{rank}(h'(\bar{u})) = n,$ (25b)

i.e.:

$$\text{span}((b(1), ..., b(k)) = I\!R^n,$$

i.e.:

$$\{v \in I\!R^n : \exists z_{\ z \in I\!R^n} \ v = z_1 b(1) + \cdots + z_p a(k)\} = I\!R^n.$$

Remark 3.1.

Our interpretation is that if $p = 0 = k$ then (A), (B), and (C) all hold; if $k = 0$ then (A) is equivalent to (B); and if $p = 0$ then (A) is equivalent to (C).

(A. Mixed) Alternative (23a) is similar to Mangasarian's generalization[16] of the Arrow-Hurwicz-Uzawa Constraint Qualification.[17] The "obvious" alternative (23b) is new. We will see that the Jacobian Criterion is sufficient for mixed Lagrange regularity. And we will show it is a *minimal* Jacobian condition.

For the mixed case, the Jacobian Criterion requires that either the sum of the wedge and span is "small," with the wedge lying strictly on one side of some hyperplane and the span lying in it, or else it is "large," spanning the whole space. In fact, that proves parts (a) and (b) are mutually exclusive.

(B. Inequalities) When only inequalities are present, the Jacobian Criterion requires that either the wedge generated by the $g'_I(\bar{u})$ is "small," lying strictly on one side of some hyperplane, or else it is "large," spanning the whole space. Again, that observation proves parts (a) and (b) are mutually exclusive. What the Criterion rules out are the "borderline" cases where neither is true — i.e., where whenever the spanned convex cone lies on one side of some hyperplane, it does not lie *strictly* on one side. Examples (5) above and (68) below are such examples, with $g^{1\prime}(0)$ and $g^{2\prime}(0)$ pointing in opposite directions.

(C. Equalities) For analogy with cases (A) and (B), we have broken condition (25) into two parts. Together they are equivalent to the single rank hypothesis of the classical Lagrange Multiplier Theorem: $\operatorname{rank}(h'(\bar{u})) = \min\{k, n\}$; see Corollary 1c below.

Removable constraints. In many applications, Lagrange multipliers satisfying (4) are sought as a step in finding the constrained maximizing point $\bar{u}$. If the Jacobian Criterion fails, all is not lost. For it may be that some constraint can be removed without changing the constraint set C near $\bar{u}$; and the Jacobian Criterion might be satisfied for the reduced set of constraints. In such situations (e.g. $h^1 = h^2$), theorems using the Jacobian rank conditions should be applied to the rank of a "maximally reduced" set of constraints.[18]

---

[16] Mangasarian's "modified Arrow-Hurwicz-Uzawa constraint qualification" in [33], pp. 172–173, which requires that h be C^1. His condition with its rank $= k$ restriction is equivalent to our part (a), with its rank $= k < n$ restriction (since the existence of such a ξ as in (23a) requires $k < n$); our two parts (a) and (b) together are weaker, however, than his modified Arrow-Hurwicz condition.

[17] Cf. the hypothesis of Theorem 3 in [4].

[18] This is an application of our remark earlier, page 103, that Lagrange regularity depends on the particular constraint functions g and h, and not just on the constraint set $C(g, h)$. Cf. Condition R_1 in [2], p. 8, where $g^\dagger$ represents a reduced set of constraints. (Page 162 in [3].) Cf. Remark (6.1.iii) below, page 126, for an analogue with path conditions where Lagrange regularity depends on the constraint functions, rather than on the constraint set.

4. The Jacobian criterion is sufficient

The Jacobian Criterion is sufficient for Lagrange regularity.

Theorem 1. *Let $U \subseteqq I\!R^n$ be open, and $\bar{u} \in U$.*

A) (Mixed: Equalities and inequalities) *The Jacobian Criterion (A)*[19] *is $(\bar{u}, \mathcal{G}_D(\bar{u}), \mathcal{H}_{DC}(\bar{u}), \mathcal{F}_D(\bar{u}))$-sufficient for Lagrange mixed-regularity.*[20] *In other words:*

Suppose $g\colon U \to I\!R^m$ is differentiable at $\bar{u}$, and $h\colon U \to I\!R^k$ belongs to $\mathcal{H}_{DC}(\bar{u})$. If the Jacobian Criterion (23a,b) holds for (g, h) at $\bar{u}$, then (g, h) is Lagrange mixed-regular for $(\bar{u}, \mathcal{F}_D(\bar{u}))$.

Then if $\bar{u}$ maximizes f subject to $g \geqq 0$ and $h = 0$, there exists a $\lambda \in I\!R^m$ and a $\mu \in I\!R^k$ such that:

$$f'(\bar{u})^T + \lambda^T g'(\bar{u}) + \mu^T h'(\bar{u}) = 0 \tag{26}$$

$$\lambda \geqq 0 \tag{27a}$$

$$\lambda_i = 0 \ \text{ if } i \in J. \tag{27b}$$

B) (Inequalities)[21] *The Jacobian Criterion (B)*[22] *is $(\bar{u}, \mathcal{G}_D(\bar{u}), \mathcal{F}_D(\bar{u}))$-sufficient for Lagrange inequality-regularity. In other words:*

Suppose $g\colon U \to I\!R^m$ is differentiable at $\bar{u}$. If the Jacobian Criterion (24) holds for g at $\bar{u}$, then g is Lagrange inequality-regular for $(\bar{u}, \mathcal{F}_D(\bar{u}))$.

Then if $\bar{u}$ maximizes f subject to $g \geqq 0$, there exists a $\lambda \in I\!R^m$ such that:[23]

$$f'(\bar{u})^T + \lambda^T g'(\bar{u}) = 0, \tag{28}$$

$$\lambda \geqq 0 \tag{29a}$$

$$\lambda_j = 0 \ \text{ if } j \in J. \tag{29b}$$

---

[19] Page 108(A).

[20] In particular, the smoothness conditions on the functions are satisfied if h is differentiable in a neighborhood of $\bar{u}$, and f and g are differentiable at $\bar{u}$. Sufficiency is defined in p. 106.

[21] In this case g is present and h is absent. Equivalently, we can set $h = 0$ in part (A) delete references to, and terms involving, μ or h.

[22] Page 108(B).

[23] J is defined in (7).

C) **(Equalities)**[24] *The Jacobian Criterion (C)*[25] *is* $(\bar{u}, \mathcal{H}_{DC}(\bar{u}), \mathcal{F}_D(\bar{u}))$*-sufficient for Lagrange equality-regularity,*[26] *and is* $(\bar{u}, \mathcal{H}_P(\bar{u}), \mathcal{F}_P(\bar{u}))$*-sufficient when* $k = n$*. In other words:*

Suppose that either $h \in \mathcal{H}_{DC}(\bar{u})$*, or that* h *has partial derivatives at* $\bar{u}$ *and that* $k = n$*. If the Jacobian Criterion (25) holds for* h *at* $\bar{u}$*, then* h *is Lagrange regular for* $(\bar{u}, \mathcal{F}_D(\bar{u}))$*.*

Then if $\bar{u}$ *maximizes* f *subject to* $h = 0$*, there exists a* $\mu \in I\!R^k$ *such that:*

$$f'(\bar{u})^T + \mu^T h'(\bar{u}) = 0. \tag{30}$$

Our proof of Theorem 1 begins on page 117.

We can immediately derive special cases of Theorem 1, by taking advantage of the "trivial" second parts, (23b), (24b), (25b), of the Jacobian Criterion. As an obvious special case of Theorem 1.A, for example, we have:

Corollary 1a. Mixed: Equalities and inequalities. *Let* $U \subseteqq I\!R^n$ *be open and let* $\bar{u} \in U$*. Suppose* $g \in \mathcal{G}_D(\bar{u})$ *and* $h \in \mathcal{H}_{DC}(\bar{u})$*, i.e.* $g\colon U \to I\!R^m$ *and* $h\colon U \to I\!R^k$ *are differentiable at* $\bar{u}$*, and* h *is continuous in a neighborhood of* $\bar{u}$*. Let* $g^1, \ldots, g^p$ *be the binding inequality constraints at* $\bar{u}$*. If:*

$$\operatorname{rank}\left(\begin{bmatrix} g_I'(\bar{u}) \\ h'(\bar{u}) \end{bmatrix}\right) = p + k, \tag{31}$$

then (g, h) *is Lagrange mixed-regular for* $(\bar{u}, \mathcal{F}_D(\bar{u}))$*, uniquely.*[27]

Proof. Rank condition (31) implies that $\operatorname{rank}(h'(\bar{u})) = k$, and also that $\binom{g_I'(\bar{u})}{h'(\bar{u})}$ is onto $I\!R^{p+k}$, so there are many ξ satisfying the Jacobian condition (23a), hence the corollary follows from Theorem 1.A. The rank condition also guarantees uniqueness of solutions (λ, μ) to $\binom{g_I'(\bar{u})}{h'(\bar{u})}^T \binom{\lambda}{\mu} = -f'(\bar{u})$, hence to (26) and (27). □

As an obvious special case of Theorem 1.B, we have:

Corollary 1b. Inequalities. *Let* $U \subseteqq I\!R^n$ *be open and let* $\bar{u} \in U$*. Suppose* $g \in \mathcal{G}_D(\bar{u})$*, so* $g\colon U \to I\!R^m$ *is differentiable at* $\bar{u}$*; and suppose* $g^1, \ldots, g^p$ *are the binding constraints at* $\bar{u}$*. If:*

$$\operatorname{rank}(g_I'(\bar{u})) = p, \tag{32}$$

then g *is Lagrange inequality-regular for* $(\bar{u}, \mathcal{F}_D(\bar{u}))$*, uniquely.*

---

[24] In this case h is present and g is absent. Equivalently, we can set $g = 0$ in part (A) and delete references to, and terms involving, λ or g.

[25] Page 108(C).

[26] In particular, the smoothness conditions on the functions are satisfied if h is differentiable in a neighborhood of $\bar{u}$, and f is differentiable at $\bar{u}$.

[27] Cf. [36], Theorem 3.1, p. 180.

Proof. Rank condition (32) implies that $g'_I(\bar{u})$ is onto $I\!R^p$, so there are many ξ satisfying the Jacobian condition (24a), hence the corollary follows from Theorem 1.B. The rank condition also implies that the solution λ to $g'_I(\bar{u})^T\lambda = -f'(\bar{u})$ is unique. □

As an obvious special case of Theorem 1.C, we have:

Corollary 1c. Equalities. *Let $U \subseteqq I\!R^n$ be open and let $\bar{u} \in U$. Suppose $h \in \mathcal{H}_{DC}(\bar{u})$, so $h\colon U \to I\!R^k$ is differentiable at $\bar{u}$, and continuous in some neighborhood of $\bar{u}$. If:*

$$\operatorname{rank}(h'(\bar{u})) = \min\{k, n\} \tag{33}$$

(i.e. $\operatorname{rank}(h'(\bar{u})) = k < n$ *or* $\operatorname{rank}(h'(\bar{u})) = n$*), then h is Lagrange equality-regular for $(\bar{u}, \mathcal{H}_{DC}(\bar{u}))$, uniquely.*

Proof. If $\operatorname{rank}(h'(\bar{u})) = k < n$, then (25a) holds, so the corollary follows from Theorem 1.C. If the rank equals n, then the Jacobian condition (25b) holds. In either case, the rank condition guarantees uniqueness of the solution to $h'(\bar{u})^T\mu = -f'(\bar{u})$. □

Remark 4.1 on Corollaries.

i) It is important to note that the rank conditions (31) and (32) refer to the Jacobian of only the *binding* constraints. Indeed, the conditions are weaker than the conditions that would result if we replaced $g'_I(\bar{u})$ by $g'(\bar{u})$. For if the m rows of $g'(\bar{u})$ are linearly independent, then so are the p rows of the submatrix $g'_I(\bar{u})$, a fortiori; and the converse fails.

On the other hand, conditions (31) and (32) are stronger than the conditions that would result if we replaced $p+k$ and p by $\min\{p+k, n\}$ and $\min\{p, n\}$, respectively, even though (31) and (32) imply $p + k \leqq n$ and $p \leqq n$, respectively. Indeed, the corollaries would fail under the weaker conditions. Consider, for example, this modification of the well known Kuhn-Tucker example [29], p. 484:

$$\begin{aligned} g^1(u_1, u_2) &= -u_1^3 - u_2 \\ g^2(u_1, u_2) &= u_1 \\ g^3(u_1, u_2) &= u_2. \end{aligned} \tag{34}$$

Here $C(g) = \{(0,0)\}$ and all three constraints are binding. Although $\operatorname{rank}(g'(0,0)) = 2 = n = \min\{p, n\}$, the rank is less than $3 = p$. Not only does our condition (32) fail, but so does Lagrange regularity (consider $f(u_1, u_2) = u_1$).

ii) Conditions (31), (32), and (33) are stronger than necessary, as simple examples show.[28]

[28] For example, if $g^1(u_1, u_2) = u_1 + u_2$ and $g^2 = g^1$, then $g = (g_1, g_2)$ is Lagrange regular at $(0,0)$, even though rank condition (32) fails. Cf. the comments on removable constraints in pp. 103, 109, and 126.

Remark 4.2 on Theorem 1. Although the constraint set $C(g, h)$ need not be open, we cannot drop the assumption that the domain U of g and h is open. Consider the function h defined by:

$$h(u_1, u_2) = u_1 + u_2 - 1 \tag{35}$$

on the nonnegative quadrant of $I\!R^2$. Even though h is continuous and differentiable at $(0, 1)$, and satisfies the Jacobian Criterion (25) there, h is not Lagrange equality-regular there (the function $f(u_1, u_2) = u_1$ is maximized at $(1, 0)$, but no Lagrange multiplier exists there). However, when we switch from Jacobian conditions to path conditions, the openness assumption will be dropped; see Theorem 3 and the remark following it.

(A. Mixed) This theorem has weaker hypotheses than most earlier results, in two respects: the Jacobian Criterion is a weaker constraint qualification (allowing the new "obvious" alternative (23b)), and the differentiability conditions on the equality constraints h^i are weaker smoothness assumptions.[29]

The continuity hypothesis on h can be weakened significantly, as indicated in the Continuity remark and Theorem 1^+ below, but it cannot be dispensed with altogether, as shown by the example in that remark.

(B. Inequalities) Part (B) solves a traditional problem, considered in Karush [25], in Kuhn and Tucker [29], and in Arrow, Hurwicz, and Uzawa [4]. The Jacobian constraint qualification (24) is weaker than the Jacobian constraint qualifications in previous work (allowing the new "obvious" alternative, (b)).

(C. Equalities) This is the classical Lagrange Multiplier Theorem, except we have weakened the usual C^1 hypothesis on h to differentiability at $\bar{u}$ and continuity in a neighborhood.[30] This allows, for example, $h = 0$ as an equality constraint, where:

$$h(u_1, u_2) = \begin{cases} u_2 - u_1^2 \sin(1/u_1), & \text{if } u_1 \neq 0 \\ \\ 0 & \text{otherwise,} \end{cases} \tag{36}$$

As another example, consider:

$$\begin{aligned} g^1(x_1, x_2) &= 6 - 3x_1 - 3x_2 \geqq 0 \\ g^2(x_1, x_2) &= 6 - 4x_1 - 2x_2 \geqq 0 \\ g^3(x_1, x_2) &= 6 - 2x_1 - 4x_2 \geqq 0, \end{aligned}$$

which are binding at $(2, 2)$, with $\operatorname{rank}(g_I'(2, 2)) < 3 = p$; so property (32) fails. Yet $g_I'(2, 2)\xi > 0$ for $\xi = (1, 1)$, so g is Lagrange regular at $(2, 2)$ by Theorem 1.B.

[29] The theorem holds for $\mathcal{H}_{DC}(\bar{u})$ instead of just for $\mathcal{H}_{C^1}(\bar{u})$. Halkin [16] used similarly weak conditions on g and h, obtaining Lagrange *quasi*-regularity as a somewhat weaker conclusion; see the Historical Comments and Comparisons in Section 8..

[30] In particular, the theorem holds for constraints in $\mathcal{H}_{DC}(\bar{u})$ instead of just for those in $\mathcal{H}_{C^1}(\bar{u})$). See the preceding footnote.

even though it is not C^1 at the origin.
When $k \leqq n$, the Lagrange multiplier vector μ of (30) is unique.[31]

Remark 4.3 on continuity of h.
i) The differentiability and continuity condition $h \in \mathcal{H}_{DC}$ in the Mixed and in the Equalities Problem is weaker than usual (cf. footnote 29). Note that when $k = n$ neither differentiability nor continuity of h is required in Theorem 1.C, but only the existence of partial derivatives. For the general case, however, continuity cannot be dropped completely, as the following "nonsubstitution" Example 4.1 shows.

Example 4.1. The function $h\colon \mathbb{R}^2 \to \mathbb{R}$:

$$h(x,y) = \begin{cases} x+y, & \text{if } x+y \neq 0 \\ x^2+y^2, & \text{if } x+y=0 \end{cases} \tag{37}$$

is differentiable at $(0,0)$, with $h'(0,0) = (1,1)$. And it satisfies the Jacobian rank condition (25a). Yet it is not Lagrange regular for even the class of linear functions, even though it is differentiable at the origin. For its zero level set allows no "substitution," since the constraint set $C(h)$ contains only the origin $(0,0)$; so if we define the function $f\colon \mathbb{R} \to \mathbb{R}$ by $f(x,y) = y$, then f is maximized on $C(h)$ at $(0,0)$, but no real μ can satisfy:

$$\begin{aligned} 0+\mu &= \frac{\partial f}{\partial x}(0,0) + \mu\frac{\partial h}{\partial x}(0,0) = 0 \\ 1+\mu &= \frac{\partial f}{\partial y}(0,0) + \mu\frac{\partial h}{\partial y}(0,0) = 0. \end{aligned} \tag{38}$$

The hypothesis of Theorem 1.C that fails here is continuity: for every x, the function h is discontinuous at $h(x,y)$ when $y = -x \neq 0$, hence h is discontinuous in every neighborhood of $(0,0)$.

ii) We can, however, weaken the continuity hypothesis on h significantly. When the Jacobian Condition holds, so $h'(\bar{u})$ has full rank k, we can partition $\mathbb{R}^n$ so that $\bar{u} = (\bar{x},\bar{y}) \in X \times Y = \mathbb{R}^{n-k} \times \mathbb{R}^k$, with $\operatorname{rank}(h_y(\bar{x},\bar{y})) = k$. Then as the proof below of Theorem 1 below makes clear, it is only continuity on Y that is needed. Indeed, the proof shows that continuity on Y is only required for a sequence of x^j converging to $\bar{x}$, and only in a certain neighborhood whose diameter diminishes with $\|x^j - \bar{x}\|$:

Theorem 1$^+$. *Let $U \subseteqq \mathbb{R}^n$ be open, and $\bar{u} \in U$.*

[31] Cf. Carathéodory [9], p. 177, Bliss [6], p. 210 for $k < n$. When $k \leqq n$, rank condition (25) guarantees that the k rows of the matrix corresponding to $h'(\bar{u})$ are linearly independent, so (30) can be uniquely solved for μ.

A) (Mixed: Equalities and inequalities) *Suppose* $g\colon U \to I\!R^m$ *and* $h\colon U \to I\!R^k$ *are differentiable at* $\bar{u}$*, and one of the following two mutually exclusive conditions (a), (b) holds:*

a) there exists a factoring $(X, Y) = I\!R^{n-k} \times I\!R^k$ *and a real* γ *with:*

$$\operatorname{rank}(h_y(\bar{x}, \bar{y})) = k < n, \tag{39a.i}$$

$$\|(h_y(\bar{x}, \bar{y}))^{-1} h_x(\bar{x}, \bar{y})\| < \gamma, \tag{39a.ii}$$

and for any $x \in X$*:*
there exist real $t^j \searrow 0$ *for which* $h(t^j x, \cdot)$ *is continuous on* $B_{\gamma\|t^j x\|}(\bar{y})$*, and there exists a* $\xi \in I\!R^n$ *such that:*

$$g'_I(\bar{u})\xi > 0,$$
$$h'(\bar{u})\xi = 0;$$

b) $\operatorname{wedge}(a(1), \ldots, a(p)) + \operatorname{span}(b(1), \ldots, b(k)) = I\!R^n$ (39b)

i.e.:

$$\{v \in I\!R^n : \exists t_{\,0 \leqq t \in I\!R^p}\ \exists z_{\,z \in I\!R^k}\ v = t_1 a(1) + \cdots + t_p a(p) + z_1 b(1) + \cdots + z_k b(k)\} = I\!R^n.$$

Then (g, h) *is Lagrange mixed-regular for* $(\bar{u}, \mathcal{F}_D(\bar{u}))$*. Then if* $\bar{u}$ *maximizes* f *subject to* $g \geqq 0$ *and* $h = 0$*, there exists a* $\lambda \in I\!R^m$ *and a* $\mu \in I\!R^k$ *such that:*

$$f'(\bar{u})^T + \lambda^T g'(\bar{u}) + \mu^T h'(\bar{u}) = 0, \tag{40a}$$

$$\lambda \geqq 0, \tag{40b}$$

$$\lambda_i = 0 \ \text{ if } i \in J. \tag{40c}$$

B) (Inequalities)[32] *This is the same as in Theorem 1(B): The Jacobian Criterion (B)*[33] *is* $(\bar{u}, \mathcal{G}_D(\bar{u}), \mathcal{F}_D(\bar{u}))$*-sufficient for Lagrange inequality-regularity.*

C) (Equalities)[34] *Suppose* $h\colon U \to I\!R^k$ *is differentiable at* $\bar{u}$*, and one of the following two mutually exclusive conditions (a), (b) holds:*

a) there exists a factoring $(X, Y) = I\!R^{n-k} \times I\!R^k$ *and a real* γ *with:*

$$\operatorname{rank}(h_y(\bar{x}, \bar{y})) = k < n, \tag{41a.i}$$

$$\|(h_y(\bar{x}, \bar{y}))^{-1} h_x(\bar{x}, \bar{y})\| < \gamma, \tag{41a.ii}$$

---

[32] In this case g is present and h is absent. Equivalently, we can set $h = 0$ in part (A) and delete references to, and terms involving, μ or h.

[33] Page 108(B).

[34] In this case h is present and g is absent. Equivalently, we can set $g = 0$ in part (A) and delete references to, and terms involving, λ or g.

and for any $x \in X$*:*
there exist real $t^j \searrow 0$ *for which* $h(t^j x, \cdot)$ *is continuous on* $B_{\gamma\|t^j x\|}(\bar{y})$*;*
b) $\operatorname{rank}(h_y(\bar{x}, \bar{y})) = k = n.$ (41b)
Then h *is Lagrange regular for* $(\bar{u}, \mathcal{F}_D)$*. So if* $\bar{u}$ *maximizes* f *subject to* $h = 0$*, there exists a* $\mu \in I\!R^k$ *such that:*

$$f'(\bar{u})^T + \mu^T h'(\bar{u}) = 0. \tag{42}$$

Remark 4.4 on continuity. i) *(Weakening continuity)* Theorem 1 used the same continuity assumption on the equality constraint functions h as Halkin [16], but obtained a stronger conclusion. Theorem 1^+, on the other hand, weakens the continuity assumption on h while retaining our full regularity conclusion. In the following example, Theorem 1^+ guarantees Lagrange regularity, even though it allows some discontinuity in every neighborhood of the constrained maximum.

Example 4.2. For the open disk $U = \{(x, y) \in I\!R^2 \colon x^2 + y^2 < 1\}$, we define the function $h \colon U \to I\!R$ by:

$$h(x, y) = \begin{cases} x + y, & \text{if } x \neq 0 \\ y + y^2, & \text{otherwise.} \end{cases} \tag{43}$$

This constraint function h satisfies the conditions of Theorem 1^+, but not those of Theorem 1. First, it is easy to see that h is differentiable at $(0, 0)$. Second, $h(x, y) = 0$ if and only if $x + y = 0$, and for any such x and y we have $\partial h(x, y)/\partial x = 1 = \partial h(x, y)/\partial y$. Third, constrained maxima of $f(x, y)$ clearly correspond to unconstrained maxima of $f(x, -x)$, and at such points we have $\partial f(x, y)/\partial x = \partial f(x, y)/\partial y$. Thus $\mu = -1$ satisfies the Lagrange equality (42) at any maximizing point (x, y).

ii) *(But not too far)* It is tempting to weaken further the continuity condition on h, by changing the strict inequalities (39a.ii) and (41.aii) to weak inequalities. That would only require continuity of $h(x, \cdot)$ on $B_{\|t^j x\|}(\bar{y})$. But our next example shows that the Lagrange regularity result of Theorem 1^+ would fail.

Example 4.3. First we define the function $g \colon I\!R \to I\!R$ by:

$$g(x) = -(x + x^3), \tag{44}$$

and then we define the function $h \colon I\!R^2 \to I\!R$ by:

$$h(x, y) = \begin{cases} x - g^{-1}(y), & \text{if } x - g^{-1}(y) \neq 0 \\ x^2 + y^2, & \text{otherwise.} \end{cases} \tag{45}$$

The function h is differentiable at the origin, and the Jacobian Criterion (25) holds, since:

$$\begin{aligned}\frac{\partial h}{\partial x}(0,0) &= 1 \\ \frac{\partial h}{\partial y}(0,0) &= 1.\end{aligned} \tag{46}$$

And $\|(h_y(0,0))^{-1}h_x(0,0)\| = 1$.

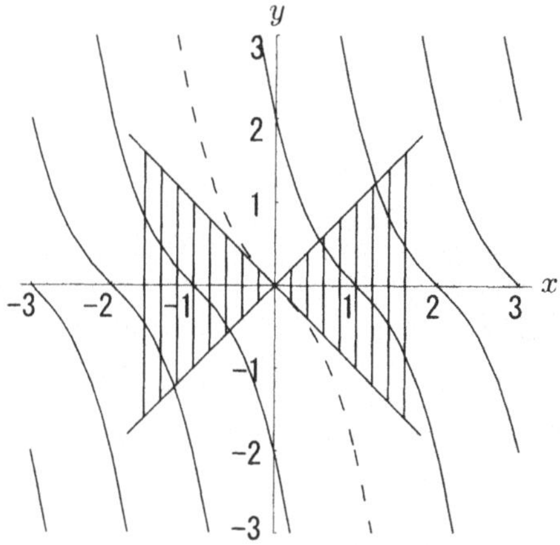

Figure 1

Here the dashed level curve through the origin is the locus of the only discontinuities of h. So for each x, the function $h(x,\,\cdot\,)$ is continuous on the y-ball $B^{K^k}_{\gamma\|x\|}(0)$ for $\gamma = \|(h_y(0,0))^{-1}h_x(0,0)\|$, though not for any greater γ.

And h is not Lagrange regular at $(0,0)$, since the constraint set $C(h)$ is the singleton origin, and the function $f(x,y) = y$ is maximized there, but cannot satisfy the regularity requirement — which is the same as in (38) in Example 4.1. The hypothesis that fails is $h(x,\,\cdot\,)$-continuity for some $\gamma > \|(h_y(0,0))^{-1}h(_x(0,0)\|$.

The proof itself of Theorem 1^+ fails because, as this same example shows, the Non-C^1 Implicit Function Theorem (Theorem 6 below) no longer holds under this weakened continuity hypothesis on h.

Proof of Theorem 1.

(A. Mixed) If (23b) holds, it is immediate that λ and μ can be found satisfying (26) and (27). So we assume that condition (23a) holds. Thus the index set[35]

[35] I is defined in (7).

$I = \{1, \dots, p\}$ of binding constraints is nonempty, and $\text{rank}(h'(\bar{u})) = k < n$. By the Reduction principle (16), it suffices to find $\lambda \in I\!R^p$ and $\mu \in I\!R^k$ satisfying (26) and (27a).

We complete the proof by contradiction. Suppose that, for some $f \in \mathcal{F}_D(\bar{u})$, there do not exist such nonnegative $\lambda \in I\!R^p$ and a $\mu \in I\!R^k$. Then by the Fundamental Lemma there exists a $\hat{z} \in I\!R^n$ such that:

$$g'_I(\bar{u})\hat{z} \geqq 0 \tag{47a}$$
$$h'(\bar{u})\hat{z} = 0 \tag{47b}$$
$$f'(\bar{u})\hat{z} > 0. \tag{47c}$$

For all $s \in (0, 1]$ define:

$$z(s) = \hat{z} + s\xi, \tag{48}$$

for any ξ satisfying (23a). With (47a,b,c), and (23a) we then have:

$$g'_I(\bar{u})z(s) > 0 \qquad \text{for all } s \in (0, 1] \tag{49}$$
$$h'(\bar{u})z(s) = 0 \qquad \text{for all } s \in (0, 1] \tag{50}$$
$$f'(\bar{u}) \cdot z(s) = f'(\bar{u}) \cdot \hat{z} + sf'(\bar{u}) \cdot \xi > 0, \tag{51}$$

for all small positive s. We define $\tilde{z} = z(s)$ for one such small s.

By the Jacobian Criterion (23a), there is a factoring $I\!R^n = X \times Y$ such that $\bar{u} = (\bar{x}, \bar{y})$ and $\text{rank}(h'(\bar{u})) = \text{rank}(h_y(\bar{x}, \bar{y})) = k$. Then from $h \in \mathcal{H}_{DC}$ it follows that, for any $x \in X$ and for any small enough $t^j \searrow 0$ the function $h(\bar{x} + t^j x, \cdot)$ is continuous on some neighborhood $B^Y(\bar{y}_i)$ of $\bar{y}$.

Applying (133) to (49), (50), and (51) with $\tilde{z} = (\tilde{x}, \tilde{y})$:

$$g_{I_u}(\bar{u})\tilde{z} = g_{I_x}(\bar{x}, \bar{y})\tilde{x} + g_{I_y}(\bar{x}, \bar{y})\tilde{y} > 0 \tag{52a}$$
$$h_u(\bar{u})\tilde{z} = h_x(\bar{x}, \bar{y})\tilde{x} + h_y(\bar{x}, \bar{y})\tilde{y} = 0 \tag{52b}$$
$$f_u(\bar{u}) \cdot \tilde{z} = f_x(\bar{x}, \bar{y})\tilde{x} + f_y(\bar{x}, \bar{y})\tilde{y} > 0. \tag{52c}$$

By the Non-C^1 Implicit Function Theorem[36] there is an $\varepsilon > 0$, a subset $X_0 = B_\varepsilon(\bar{x}) \cap \{t^j\tilde{x} \colon j = 1, 2, 3 \dots\} \subseteqq I\!R^{n-k}$, and a function $y(\cdot) \colon X_0 \to I\!R^k$ that is differentiable at $\bar{x}$, such that:

$$y(\bar{x}) = \bar{y} \tag{53a}$$
$$h(x, y(x)) = 0 \qquad \text{for all } x \in X_0. \tag{53b}$$
$$y'(\bar{x}) = -(h_y(\bar{x}, \bar{y}))^{-1} h_x(\bar{x}, \bar{y}). \tag{53c}$$

By (52b) we also have:

---

[36] See the Appendix.

$$\tilde{y} = -(h_y(\bar{x}, \bar{y}))^{-1} h_x(\bar{x}, \bar{y})\tilde{x}, \tag{54}$$

so by the Chain Rule, (53c), and (54):[37]

$$\frac{d}{dt}|_{t=0}\; y(\bar{x} + t\tilde{x}) = y'(\bar{x})\tilde{x} = \tilde{y}. \tag{55}$$

Then the Chain Rule, (52a), and (52c) imply:

$$\frac{d}{dt}|_{t=0}\; g_I(\bar{x} + t\tilde{x}, y(\bar{x} + t\tilde{x})) = g_{I_x}(\bar{x}, \bar{y}) \cdot \tilde{x} + g_{I_y}(\bar{x}, \bar{y}) \cdot \tilde{y} > 0 \tag{56a}$$

$$\frac{d}{dt}|_{t=0}\; f(\bar{x} + t\tilde{x}, y(\bar{x} + t\tilde{x})) = f_x(\bar{x}, \bar{y}) \cdot \tilde{x} + f_y(\bar{x}, \bar{y}) \cdot \tilde{y} > 0, \tag{56b}$$

so for all small $t^j > 0$,

$$g_I(\bar{x} + t^j\tilde{x}, y(\bar{x} + t^j\tilde{x})) > 0 \tag{57a}$$

$$g_J(\bar{x} + t^j\tilde{x}, y(\bar{x} + t^j\tilde{x})) \geqq 0 \qquad \text{(by (7))} \tag{57b}$$

$$f(\bar{x} + t^j\tilde{x}, y(\bar{x} + t^j\tilde{x})) > f(\bar{x}, \bar{y}) \tag{57c}$$

$$h(\bar{x} + t^j\tilde{x}, y(\bar{x} + t^j\tilde{x})) = 0 \qquad \text{(by (53b))}. \tag{57d}$$

So for small enough $t^j > 0$ there are points $(\bar{x} + t^j\tilde{x}, y(\bar{x} + t^j\tilde{x})) \in C(g, h)$ at which f attains a value higher than $f(\bar{x}, \bar{y})$, contradicting that $\bar{u} = (\bar{x}, \bar{y})$ maximizes f, and concluding our proof for part (A).

Proofs for parts (B) and (C) can be constructed along similar lines, but here are sketches.

(B. Inequalities) If (24b) holds, then 0 belongs to the wedge generated by the rows of $g_I'(\bar{u})$, so it is again immediate that λ can be found satisfying (28) and (29). So we assume that (24a) holds: $g_I'(\bar{u})\xi > 0$ for some ξ.

If no solution exists then by the Fundamental Lemma there exists a $\hat{z} \in I\!R^n$ satisfying (47a) and (47c). For all $s \in (0, 1]$ define $z(s)$ as in (48), so (49) and (51) hold. Thus for small positive s values, $z(s)$ belongs to the constraint set $C(g)$, and also increases the maximand above $f(\bar{u})$, contradicting the constrained maximality of $f(\bar{u})$.

(C. Equalities) We repeat the proof for the mixed part (A), replacing ξ by $\hat{z}$ and omitting all statements concerning the inequality constraints g: (47a), (49), (52a), (56a), and (57a,b). □

Proof of Theorem 1^+. An almost identical proof applies. The only change is to choose any $\gamma > \|(h_y(\bar{x}, \bar{y}))^{-1} h_x(\bar{x}, \bar{y})\|$, and then replace the single neighborhood $B^Y(\bar{y})$ by the neighborhoods $B^Y_{\gamma\|t^j x\|}(\bar{y})$ of radii $\gamma\|t^j x\|$. □

---

[37] The derivative is with respect to $t \in \{t^j : j = 1, 2, 3, \dots\}$ (our implicit function theorem only guaranteed that $y(\bar{x} + t^j\tilde{x})$ is defined for such t). It is easy to verify that the Chain Rule does apply in this context.

5. The Jacobian criterion is minimal

Section 4. showed the Jacobian Criterion is sufficient for Lagrange regularity. Example 6.1 below, p. 124, shows it is not necessary. Indeed, it shows that no Jacobian constraint qualification can be both necessary and sufficient for Lagrange regularity. Jacobian conditions are blunt instruments for detecting Lagrange regularity.[38] Nevertheless, Jacobian conditions are of special interest, because they typically have computability properties useful in practical applications.[39] So it is important to show that the Jacobian Criterion of the previous sections is as weak as possible among Jacobian conditions: if one restricts oneself to Jacobian conditions sufficient for Lagrange regularity, the Jacobian Criterion is minimal (over an appropriate class of constraint functions).

In contrast with sufficiency theorems, weaker smoothness hypotheses on the constraint functions do not necessarily make better minimality theorems. To prove minimality of a Jacobian condition for classes $\mathcal{G}$ and $\mathcal{H}$ of inequality and equality constraints, we must show that failure of the condition for some g and h implies there exist $\hat{g}$ and $\hat{h}$ *in the same classes* $\mathcal{G}$ and $\mathcal{H}$, respectively, that are not Lagrange regular. The smaller the classes $\mathcal{G}$ and $\mathcal{H}$, the harder it may be to obtain such $\hat{g}$ and $\hat{h}$. (And there is no guarantee that every class $\mathcal{H}$ will have a minimally sufficient Jacobian condition for Lagrange regularity.[40]) We present below a minimality result for any classes $\mathcal{G}$ and $\mathcal{H}$ that contain at least the quadratic functions.

Theorem 2. *Let U be an open subset of $\mathbb{R}^n$, let $\bar{u} \in U \subseteqq \mathbb{R}^{n-k} \times \mathbb{R}^k$, and suppose that function classes $\mathcal{G} \subseteqq \mathcal{G}_D(\bar{u})$ and $\mathcal{H} \subseteqq \mathcal{H}_D(\bar{u})$ contain all quadratic functions.*[41]

- **A) (Mixed: Equalities and inequalities)** *If the Jacobian Criterion (A) is $(\bar{u}, \mathcal{G}, \mathcal{H}, \mathcal{F}_D(\bar{u}))$-sufficient for Lagrange mixed-regularity, then it is minimally $(\bar{u}, \mathcal{G}, \mathcal{H}, \mathcal{F}_D(\bar{u}))$-sufficient for Lagrange mixed-regularity.*
- **B) (Inequalities)** *If the Jacobian Criterion (B) is $(\bar{u}, \mathcal{G}, \mathcal{F}_D(\bar{u}))$-sufficient for Lagrange inequality-regularity, then it is minimally $(\bar{u}, \mathcal{G}, \mathcal{F}_D(\bar{u}))$-sufficient for Lagrange inequality-regularity.*

---

[38] Although we obtain a necessary and sufficient constraint qualification in Section 6. (Theorems 3 and 4) below, it is a "path condition," not a Jacobian condition.

[39] By contrast, path conditions (Section 6) may be difficult to apply.

[40] An extreme example is any class $\mathcal{H}$ consisting of a singleton h. If there is a Jacobian condition $\mathbb{Q}$ that is $(\bar{u}, \mathcal{H}, \mathcal{F}_D(\bar{u}))$-sufficient for regularity, then any $\hat{h}$ with $\hat{h}'(\bar{u}) \notin Q$ must have $\hat{h}'(\bar{u}) \neq h'(\bar{u})$, and so cannot belong to $\mathcal{H}$.

[41] See p. 102 for definitions of Lagrange regularity, and p. 106 for definitions of minimal sufficiency.

In fact, as seen from the proof below, Theorem 2 remains true even if the class $\mathcal{F}$ of maximands is as narrow as the class of linear functions.

C) (Equalities) *If the Jacobian Criterion (C) is* $(\bar{u}, \mathcal{H}, \mathcal{F}_D(\bar{u}))$*-sufficient for Lagrange equality-regularity, then it is minimally* $(\bar{u}, \mathcal{H}, \mathcal{F}_D(\bar{u}))$*-sufficient for Lagrange equality-regularity.*

In particular, therefore, by Theorem 1 we have this corollary for $\mathcal{G} = \mathcal{G}_D$ and $\mathcal{H} = \mathcal{H}_{DC}$:

Corollary of Theorem 2. *Let* U *be an open subset of* $I\!R^n$ *and let* $\bar{u} = (\bar{x}, \bar{y}) \in U \subseteqq I\!R^n$.

A) (Mixed: Equalities and inequalities) *The Jacobian Criterion (A) is minimally* $(\bar{u}, \mathcal{G}_D(\bar{u}), \mathcal{H}_{DC}(\bar{u}), \mathcal{F}_D(\bar{u}))$*-sufficient for Lagrange mixed-regularity.*

B) (Inequalities) *The Jacobian Criterion (B) is minimally* $(\bar{u}, \mathcal{G}_D(\bar{u}), \mathcal{F}_D(\bar{u}))$*-sufficient for Lagrange inequality-regularity.*

C) (Equalities) *The Jacobian Criterion (C),*

$$\operatorname{rank}(h'(\bar{u})) = \min\{k, n\}, \tag{58}$$

is minimally $(\bar{u}, \mathcal{H}_{DC}(\bar{u}), \mathcal{F}_D(\bar{u}))$*-sufficient for Lagrange equality-regularity.*

Proof of Theorem 2.

(Part (A)) Let the Jacobian Criterion (A) be $(\bar{u}, \mathcal{G}, \mathcal{H}, \mathcal{F}_D(\bar{u}))$-sufficient for Lagrange mixed-regularity, so property (i) of the definition of minimal sufficiency[42] holds.

To see that part (ii) of the minimality definition holds, suppose some $g \in \mathcal{G}$ and $h \in \mathcal{H}$ do not satisfy the Jacobian Criterion (A). Then $p > 0$ or $k > 0$ by Remark 3.1; in what follows, references to g should be disregarded when $p = 0$, and references to h should be disregarded when $k = 0$. It then suffices to show there exist $\hat{g} \in \mathcal{G}$ and $\hat{h} \in \mathcal{H}$ with $\hat{g}_I'(\bar{u}) = g_I'(\bar{u})$ and $\hat{h}'(\bar{u}) = h'(\bar{u})$, for which $(\hat{g}, \hat{h})$ is not Lagrange regular for $(\bar{u}, \mathcal{F}_D(\bar{u}))$.

Without loss of generality we assume U is a neighborhood of the origin $0 \in I\!R^n$, and that $\bar{u} = 0$. Define matrix $A = g_I'(0)$ and matrix $B = h'(0)$. Since (g, h) violates the Jacobian Criterion (A), we know that (A, B) cannot satisfy either of these two mutually exclusive conditions:

a) $\operatorname{rank}(B) = k < n$ and there exists a $\xi \in I\!R^n$ such that: (59a)

$$A\xi > 0$$
$$B\xi = 0.$$

[42] Page 106

b) $\text{wedge}(\{a(1),\ldots,a(p)\}) + \text{span}(\{b(1),\ldots,b(k)\}) = I\!R^n$, (59b)
i.e.:

$$\{t_1 a(1) + \cdots + t_m a(p) + z_1 b(1) + \cdots + z_k b(k) : \\ \text{real } t_1,\ldots,t_p \geqq 0 \ \& \ z_1,\ldots,z_k \in I\!R\} = I\!R^n.$$

It will suffice to find $\hat{g}$ and $\hat{h}$ with $\hat{g}'_I(\bar{u}) = g'_I(\bar{u})$ and $\hat{h}'(\bar{u}) = h'(\bar{u})$, and such that $C(\hat{g},\hat{h}) = \{0\}$. For then $(\hat{g},\hat{h})$ is not Lagrange regular, because the failure of (59b) implies that there is some $\gamma \in I\!R^n$ with:

$$\gamma \notin \text{wedge}(\{a(1),\ldots,a(p)\}) + \text{span}(\{b(1),\ldots,b(k)\}); \tag{60}$$

so if we define $f(x) = -\gamma \cdot x$ then f is maximized on $C(\hat{g},\hat{h})$ only at the origin 0, and $-f'(0) = \gamma$, which by (60) cannot satisfy (9) and (10) as required by Lagrange regularity of $(\hat{g},\hat{h})$.

To find such $\hat{g}$ and $\hat{h}$, first consider the special case that some row $a(i)$ of A equals $0 \in I\!R^n$. Then for $x \in I\!R^n$ define $\hat{g}^i(x) = -(x_1^2 + \cdots + x_n^2)$, so $a(i) = \hat{g}^{i\prime}(0)$, and $\hat{g}^i(x) \geqq 0$ if and only if $x = 0$; define $\hat{g}^j = g^j$ for all $j \neq i$. Since $\mathcal{G}$ contains all quadratic functions, $\hat{g} \in \mathcal{G}$; also $\hat{g}'_I(0) = A$, and $C(\hat{g},h) = \{0\}$. Since $\hat{g}^i$ is clearly binding at $\bar{u} = 0$, the same functions in $\hat{g}$ as in g are binding at $\bar{u} = 0$. So $\hat{g}_I = g_I$. We can then define $\hat{h} = h$, and the proof is finished for the special case when some row $a(i) = 0$.

Similarly, if some row $b(j) = 0 \in I\!R^n$, define $\hat{g} = g$ and $\hat{h}^j(x_1,\ldots,x_n) = -(x_1^2 + \cdots + x_n^2)$ and $\hat{h}^i = h^i$ for all $i \neq j$; by a similar argument, $\hat{h} \in \mathcal{H}$, $\hat{h}'(0) = B$, and $C(g,\hat{h}) = \{0\}$. Since $\hat{g}_I = g_I$ and $\hat{h}'(0) = h'(0)$, the proof is finished for the special case when some row $b(j) = 0$.

It remains to consider the case that:

$$\begin{aligned} a(i) &\neq 0 \qquad \text{for all } i = 1,\ldots,p \\ b(j) &\neq 0 \qquad \text{for all } j = 1,\ldots,k. \end{aligned} \tag{61}$$

We noted that (59a) fails. There are two ways it can fail: i) we might have $\text{rank}(B) < k < n$, or ii) there might exist no ξ with $A\xi > 0$ and $B\xi = 0$.

(i) First consider the case that $\text{rank}(B) < k$. Then (61) implies $\text{rank}(B) > 0$, hence $k > 1$. Then some row of B is a linear combination of the others, say:

$$b(1) = z_2 b(2) + \cdots + z_k b(k) \qquad \text{for some real } z_2,\ldots,z_k. \tag{62}$$

By (61) we have $b(1) \neq 0$, so without loss of generality we can choose a basis so that:

$$b(1) = (1,0,\ldots,0). \tag{63}$$

Now define:

$$\begin{aligned} \hat{g}^i(x_1,\ldots,x_n) &= a(i)\cdot x \qquad \text{for all } i = 1,\ldots,p \\ \hat{h}^1(x_1,\ldots,x_n) &= x_1 - (x_2^2 + \cdots + x_n^2) \\ \hat{h}^i(x_1,\ldots,x_n) &= b(i)\cdot x \qquad \text{for all } i = 2,\ldots,k. \end{aligned} \tag{64}$$

Then $\hat{g}'_I(0) = A = g'_I(0)$ and $\hat{h}'(0) = B = h'(0)$, $\hat{g} \in \mathcal{G}$ and $\hat{h} \in \mathcal{H}$; and if $\hat{g}(x) \geqq 0$ & $\hat{h}(x) = 0$ it follows from (62), (63), and (64) that $x_1 = \cdots = x_n = 0$, so again the constraint set contains just the origin: $C(\hat{g},\hat{h}) = \{0\}$, and we are done for case (i).

(ii) Property (59a) could also fail by absence of $\xi \in I\!R^n$ with $A\xi > 0$ and $B\xi = 0$. Then the Theorem of the Alternative,[43] yields $t \in I\!R^m$, $v \in I\!R^k$ with:

$$\begin{aligned} t &\geq 0 \\ t^T A + v^T B &= 0 \in I\!R^n. \end{aligned} \tag{65a}$$

Without loss of generality we can rewrite this as:

$$a(1) = -(w_2 a(2) + \cdots + w_m a(p) + z_1 b(1) + \cdots + z_k b(k)), \tag{65b}$$

for some real z_i and some real $w_j \geqq 0$. Because of (61) we can, also without loss of generality, choose a basis so that:

$$a(1) = (1, 0, \ldots, 0) \in I\!R^n. \tag{66}$$

Now define:

$$\begin{aligned} \hat{g}^1(x_1,\ldots,x_n) &= x_1 - (x_2^2 + \cdots + x_n^2) \\ \hat{g}^i(x_1,\ldots,x_n) &= a(i)\cdot x \qquad \text{for all } i = 2,\ldots,p \\ \hat{h}^i(x_1,\ldots,x_n) &= b(i)\cdot x \qquad \text{for all } i = 1,\ldots,k. \end{aligned} \tag{67}$$

Since $\hat{g}^1$ is clearly binding at $\bar{u} = 0$, it follows that $\hat{g}'_I(0) = A = g'_I(0)$ and $\hat{h}'(0) = B = h'(0)$, $\hat{g} \in \mathcal{G}$ and $\hat{h} \in \mathcal{H}$, and if $\hat{g}(x) \geqq 0$ & $\hat{h}(x) = 0$ it follows from (65), (66), and (67) that $x_1 = \cdots = x_n = 0$; so again the constraint set contains just the origin: $C(\hat{g},\hat{h}) = \{0\}$, and we are done for this last case (ii).

At this point we have shown for part (A) that $C(\hat{g},\hat{h}) = \{0\}$, which by the sufficiency argument following (59b) completes the proof for part (A).

(Parts (B) and (C)) These follow from obvious modifications of our proof for Part (A), or as a corollary of that part if we set $h = 0$ or $g = 0$.

□

[43] See the Appendix, applying the special case of part (II) when B and β are absent.

6. The tangency-path criterion

Why consider more than the Jacobian? Because the Lagrange multiplier property (9) is a local property, one might expect that Jacobian properties would suffice to characterize conditions under which Lagrange regularity holds for (g, h). However, they cannot determine Lagrange regularity in all cases, as we see from the following example.

Example 6.1. Consider the inequalities $g \geqq 0$ where:

$$g^1(u_1, u_2) = \begin{cases} u_2 - u_1^2 \sin(1/u_1), & \text{if } u_1 \neq 0 \\ 0 & \text{otherwise} \end{cases} \tag{68}$$
$$g^2(u_1, u_2) = -u_2.$$

It will be evident from Theorem 3 that, for any partially differentiable function f that is maximized at $\bar{u} = (0, 0)$ subject to the constraints (68), there do exist Lagrange multipliers satisfying (9) and (10). By contrast, the function g of (5) is not Lagrange inequality-regular for even the class of linear functions, even though it has the same Jacobian $g'(0, 0)$ as (68) above.[44] Similar remarks apply to equality constraints.

Ensuring Lagrange regularity amounts to guaranteeing existence of feasible paths whose tangents have certain algebraic properties (see our discussion of (21) in Part 2.2). Such paths were provided in our proof Theorem 1 through Jacobian conditions on g and h. Now, in the spirit of Karush, Kuhn and Tucker, and others, we will obtain them by directly assuming their existence, but with the weakest possible assumptions.

Tangency-path criterion. Suppose $\bar{u} \in U \subsetneqq I\!R^n$, and let $(g, h)\colon U \to I\!R^m \times I\!R^k$.

A) (Mixed: Equalities and inequalities) We say that (g, h) satisfies the *Mixed-Problem Tangency-Path Criterion at* $\bar{u}$ if:

i) all the g^i and h^i have partial derivatives at $\bar{u}$,

ii) for any $v \in I\!R^n$,[45]

$$\left.\begin{array}{r} {g_I}'(\bar{u})v \geqq 0 \\ h'(\bar{u})v = 0 \end{array}\right\} \Rightarrow v \in \mathrm{cl}(\mathrm{ch}(T_{\bar{u}}C(g, h))). \tag{69a}$$

[44] Cf. the discussion of (5), p. 101.

[45] I is defined in (7).

We can also write (ii) as:

$$L(g,h) \subseteqq V(g,h), \tag{69b}$$

where we define:

$$L(g,h) = \{v \in I\!R^n : g_I{}'(\bar{u})v \geqq 0 \ \ \& \ \ h'(\bar{u})v = 0\}, \tag{70}$$

and:

$$V(g,h) = \mathrm{cl}(\mathrm{ch}(T_{\bar{u}}C(g,h))). \tag{71}$$

B) (Inequalities) We say that (g,h) satisfies the *Inequalities-Problem Tangency-Path Criterion at* $\bar{u}$ if:

i) all the g^i have partial derivatives at $\bar{u}$,

ii) for any $v \in I\!R^n$,

$$g_I{}'(\bar{u})v \geqq 0 \ \ \Rightarrow \ \ v \in \mathrm{cl}(\mathrm{ch}(T_{\bar{u}}C(g))), \tag{72a}$$

i.e.

$$L(g) \subseteqq V(g) \tag{72b}$$

in notation paralleling (69b).

C) (Equalities) We say that h satisfies the *Equalities-Problem Tangency-Path Criterion at* $\bar{u}$ if:

i) all the h^i have partial derivatives at $\bar{u}$,

ii) for any $v \in I\!R^n$,

$$h'(\bar{u})v = 0 \ \ \Rightarrow \ \ v \in \mathrm{cl}(\mathrm{ch}(T_{\bar{u}}C(h))). \tag{73a}$$

i.e.

$$L(h) \subseteqq V(h) \tag{73b}$$

in notation paralleling (69b).

A path interpretation of the criterion. Paraphrased in terms of binding constraints g^i, (72a) says that if a vector v has a nonnegative inner product $g^{i\prime}(\bar{u}) \cdot v$ with each $g^{i\prime}(\bar{u})$, then v is a limit of positive linear combinations of tangents of "feasible" paths. In particular, if the g^i were differentiable[46] then any direction v in which all the g^i had positive derivatives would be, if not a direction (i.e., tangent) of a feasible path, then at least a positive linear combination of directions of such paths.

[46] Not merely possessing partial derivatives.

Remark 6.1.

i) One might consider the simpler condition obtained from replacing $\mathrm{cl}(\mathrm{ch}(T_{\bar{u}}C(g)))$, for example, by its subset $\mathrm{ch}(T_{\bar{u}}C(g))$:[47]

$$L(g) \subseteqq \mathrm{ch}(T_{\bar{u}}C(g)). \tag{74}$$

However, that would impose a stronger requirement on constraints than does (72b), as could be seen from this example:

$$\begin{aligned} g^1(x,y,z) &= z^2 - y^2 - (x - \|z\|)^2 \\ g^2(x,y,z) &= x. \end{aligned} \tag{75}$$

ii) If we replace g^1 by

$$\hat{g}^1(x,y,z) = \begin{cases} g^1(x,y,z)/\sqrt{(x^2+y^2+z^2)}, & \text{if } (x,y,z) \neq (0,0,0) \\ 0, & \text{otherwise,} \end{cases} \tag{76}$$

then $(\hat{g}^1, g^2)$ has the same partial derivatives at the origin $(0,0,0)$ as in (75), but it is no longer Fréchet differentiable at the origin, even though it defines the same constraint set as (g^1, g^2). This illustrates the fact that Theorem 3 guarantees Lagrange regularity even when full differentiability does not apply. The example still exhibits a discrepancy between $\mathrm{cl}(\mathrm{ch}(T_0C(g))$ and $\mathrm{ch}(T_0C(g))$.

iii) Without g^2, the system (75) would not be Lagrange regular, even though the constraint set would be the same. This again emphasizes that Lagrange regularity is a property of constraint functions, not just of constraint sets.[48]

We will show that the Tangency-Path Criterion is necessary and sufficient for Lagrange regularity. By the definition of the Tangency-Path Criterion, the functions g^i and h^i in the following theorem have partial derivatives; but it is not assumed that they are Fréchet, or even Gâteaux differentiable.[49]

---

[47] Cf. [13].

[48] Cf. our comments on this dependence above, page 103, and our comments on Removable Constraints above, p. 109, for Jacobian conditions. Also see [2], footnote 2, p. 2, where again the form of constraints influences whether the rank constraint qualification holds or not — indeed, whether the inequality constraint is Lagrange regular or not.

[49] Gould and Tolle [15], p. 167, presupposing differentiability and continuity conditions that are not made in Theorems 3, 4, 5, or 6 below, state a criterion that is the dual of a condition stronger than the Tangency-Path Criterion. (See Section 8. below.)

Our Theorem 3 is a strengthening of Arrow, Hurwicz, and Uzawa's Theorem 1 [4], as well as of the "if" part of Gould and Tolle's theorem. Our Theorem 4 is a strengthening of Theorem 2 in Arrow, Hurwicz, and Uzawa [4]; it plays the same role, under weaker hypotheses, as the "only if" part of Gould and Tolle's theorem.

Theorem 3. The Tangency-path criterion is sufficient for Lagrange regularity. *Let $U \subseteqq I\!R^n$, and let $\bar{u} \in U$ be a limit point.*

A) (Mixed: Equalities and inequalities) *The Mixed-Problem Tangency-Path Criterion is sufficient for Lagrange mixed-regularity for $(\bar{u}, \mathcal{F}_D(\bar{u}))$.*
In other words, if $f: U \longrightarrow I\!R$ is differentiable at $\bar{u}$, if $g: U \longrightarrow I\!R^m$, if $h: U \longrightarrow I\!R^k$, if $\bar{u}$ maximizes f subject to $g \geqq 0$ and $h = 0$, and if (g, h) satisfies the Mixed-Problem Tangency-Path Criterion at $\bar{u}$, then there exists a $\lambda \in I\!R^m$ and a $\mu \in I\!R^k$ such that:

$$f'(\bar{u})^T + \lambda^T g'(\bar{u}) + \mu^T h'(\bar{u}) = 0, \tag{77}$$

$$\lambda \geqq 0 \tag{78a}$$

$$\lambda_i = 0 \quad \text{if } i \in J. \tag{78b}$$

B) (Inequalities) *The Inequalities-Problem Tangency-Path Criterion is sufficient for Lagrange inequality-regularity for $(\bar{u}, \mathcal{F}_D(\bar{u}))$.*
In other words, if $g: U \longrightarrow I\!R^m$, if $\bar{u}$ maximizes $f: U \longrightarrow I\!R$ subject to $g \geqq 0$, if f is differentiable at $\bar{u}$, and if g satisfies the Inequalities-Problem Tangency-Path Criterion at $\bar{u}$, then there exists a $\lambda \in I\!R^m$ such that:

$$f'(\bar{u})^T + \lambda^T g'(\bar{u}) = 0, \tag{79}$$

$$\lambda \geqq 0 \tag{80a}$$

$$\lambda_j = 0 \quad \text{if } j \in J. \tag{80b}$$

C) (Equalities) *The Equalities-Problem Tangency-Path Criterion is sufficient for Lagrange equality-regularity for $(\bar{u}, \mathcal{F}_D(\bar{u}))$.*
In other words, if $h: U \longrightarrow I\!R^m$, if $\bar{u}$ maximizes $f: U \longrightarrow I\!R$ subject to $h = 0$, if f is differentiable at $\bar{u}$, and if the h satisfies the Equalities-Problem Tangency-Path Criterion at $\bar{u}$, then there exists a $\mu \in I\!R^m$ such that:

$$f'(\bar{u})^T + \mu^T h'(\bar{u}) = 0. \tag{81}$$

Remark 6.2. Theorem 3's sufficiency result does not require that the domain U be open, in contrast to Theorem 1's sufficiency result. (Also in contrast to Theorem 1's Jacobian condition (25), our Tangency-Path Criterion fails for the function h in the non-Lagrange-regular example (35), when applied to the nonnegative quadrant of $I\!R^2$.[50])

[50] This does not contradict Theorem 1.C, which required the set U to be open.

Conversion of equalities to inequalities. Before proving Theorem 3, we note that, when working with constraint qualifications in the form of path conditions, it is sometimes convenient to convert each equality $h^i = 0$ into two inequalities $\tilde{g}^{i_1} \geqq 0$ and $\tilde{g}^{i_2} \geqq 0$, where

$$\begin{aligned}\tilde{g}^{i_1} &= h^i \\ \tilde{g}^{i_2} &= -h^i.\end{aligned} \tag{82}$$

Thus from the original mixed constraints (g, h), we obtain a corresponding set $\tilde{g}(h)$ of just inequality constraints.[51] Clearly the original constraint set remains the same: $C(\tilde{g}(h)) = C(h)$. So the problem of maximizing a function f on U subject to $\tilde{g}(h) \geqq 0$ is equivalent to maximizing f on U subject to $h = 0$.

While the constrained optimization problems are equivalent for $\tilde{g}(h)$ and h, we must ask whether our constraint qualifications remain the same in passing from h to $\tilde{g}(h)$. For our Jacobian conditions (25) and (24), the answer is no.[52]

But for our path-type constraint qualifications (73) and (72), the answer is yes, as we now show.

Lemma 6.1 (Conversion lemma). *For any inequality constraint functions g and equality constraint functions h that have partial derivatives at $\bar{u} \in U$:*

a) i) The constraints (g, h) satisfy the Mixed Tangency-Path Criterion (69) if and only if the associated inequalities $\tilde{g}(h)$ satisfy the Inequalities Tangency-Path Criterion (72).

ii) The constraint h satisfy the Equality Tangency-Path Criterion (73) if and only if the associated inequalities $\tilde{g}(h)$ satisfy the Inequalities Tangency-Path Criterion (72).

b) For any function $f : U \to I\!R$ that has partial derivatives at $\bar{u}$, there exist $\lambda \geqq 0$, with $\lambda_i = 0$ if g^i is not binding, and μ such that:

$$f'(\bar{u}) + \lambda g'_I(\bar{u}) + \mu h'(\bar{u}) = 0 \tag{83}$$

if and only if there exist $\tilde{\lambda} \geqq 0$, with $\tilde{\lambda}_i = 0$ if $\tilde{g}(h)^i$ is not binding, and such that:

$$f'(\bar{u}) + \tilde{\lambda}\tilde{g}(h)'_I(\bar{u}) = 0. \tag{84}$$

Proof. (Part (a.i)) First note that, because (g, h) and $\tilde{g}(h)$ define the same constraint set $C(g, h) = C(\tilde{g}(h))$, the right hand side $\mathrm{cl}(\mathrm{ch}(T_{\bar{u}}C(g, h)))$ of (69a) is the same as the right hand side $\mathrm{cl}(\mathrm{ch}(T_{\bar{u}}C(\tilde{g}(h)))$ of (72a). So it suffices to

[51] When h is absent, then $\tilde{g}(h)$ is understood to be g, and when g is absent, then $\tilde{g}(h)$ is understood to contain the inequalities specified in (82).

[52] For example, if $\tilde{g}(h) = (\tilde{g}_1, \tilde{g}_2) = (\tilde{g}_1, -\tilde{g}_1) \leqq 0$ corresponds to $h = 0$, then clearly there can be no vector ξ such that $\tilde{g}(h)'(\bar{y})\xi > 0$.

prove that the left hand sides of (69a) and (72a) are equivalent. And for that it clearly suffices to show that, for each $i = 1, \ldots, k$,

$$h^{i\prime}(\bar{u})v = 0 \quad \Leftrightarrow$$
$$\begin{cases} \tilde{g}^{i_1\prime}(\bar{u})v \geqq 0, & \text{if } \tilde{g}^{i_1} \text{ is binding and } \tilde{g}^{i_2} \text{ is not} \\ \tilde{g}^{i_2\prime}(\bar{u})v \geqq 0, & \text{if } \tilde{g}^{i_2} \text{ is binding and } \tilde{g}^{i_1} \text{ is not} \\ \tilde{g}^{i_1\prime}(\bar{u})v \geqq 0 \ \ \& \ \ \tilde{g}^{i_2\prime}(\bar{u})v \geqq 0, & \text{if } \tilde{g}^{i_1} \text{ and } \tilde{g}^{i_2} \text{ are both binding.} \end{cases} \tag{85}$$

Suppose first that

$$h^{i\prime}(\bar{u})v = 0. \tag{86}$$

Then it is immediate from the definition of the $\tilde{g}^i$ in (82) that $\tilde{g}^j(\bar{u})'v = 0$ for both $j = 1$ and $j = 2$.

Before proving the converse, we first note that:

$$\begin{aligned} &\text{if } \tilde{g}^{i_1} \text{ or } \tilde{g}^{i_2} \text{ is not binding at } \bar{u}, \\ &\text{then } 0 = h'(\bar{u}) = \tilde{g}^{i_1\prime}(\bar{u}) = \tilde{g}^{i_2\prime}(\bar{u}). \end{aligned} \tag{87}$$

For if $\tilde{g}^{i_1}$ is not binding, then there is a neighborhood of $\bar{u}$ in which $\tilde{g}^{i_1}$ is nonnegative. Since $\tilde{g}^{i_1}(\bar{u}) = h^i(\bar{u}) = 0$, this means that $\tilde{g}^{i_1}$ is minimized at $\bar{u}$ in that neighborhood, and so $0 = \tilde{g}^{i_1\prime}(\bar{u}) = h^{i\prime}(\bar{u})$, and $\tilde{g}^{i_2}(\bar{u}) = -h^{i\prime}(\bar{u}) = 0$.

Now to prove the converse in (85), suppose that one of the two first conditions on the right hand side is satisfied. Since at least one of the $\tilde{g}^i$ is not binding, it follows from (87) that $h'(\bar{u}) = 0$, so $h^{i\prime}(\bar{u})v = 0$.

Finally, suppose the last case on the right hand side holds. By definition (82), that implies $h'(\bar{u})v \geqq 0$ and $-h'(\bar{u})v \geqq 0$, from which the left hand side of (85) follows immediately.

(Part (a.ii)) Apply an argument similar to that for (a.i).

(Part (b)) First, suppose there exist $\lambda \geqq 0$, with $\lambda_i = 0$ if g^i is not binding, and μ as in (83). For each $i = 1, \ldots, k$ with $\mu_i = 0$, we choose $\tilde{\lambda}_{i_1} = \tilde{\lambda}_{i_2} = 0$. For each $\mu_i \neq 0$, we choose any $\tilde{\lambda}_{i_1} \geqq 0$ and $\tilde{\lambda}_{i_2} \geqq 0$ such that $\tilde{\lambda}_{i_1} - \tilde{\lambda}_{i_2} = \mu_i$; then:

$$\begin{aligned} \tilde{\lambda}_{i_1}\tilde{g}(h)^{i_1\prime}(\bar{u}) + \tilde{\lambda}_{i_2}\tilde{g}(h)^{i_2\prime}(\bar{u}) &= \tilde{\lambda}_{i_1}h^{i\prime}(\bar{u}) - \tilde{\lambda}_{i_2}h^{i\prime}(\bar{u}) \\ &= (\tilde{\lambda}_{i_1} - \tilde{\lambda}_{i_2})h^{i\prime}(\bar{u}) \\ &= \mu_i h^{i\prime}(\bar{u}). \end{aligned} \tag{88}$$

With those $\tilde{\lambda}_i$, (84) is satisfied. And we can further choose $\tilde{\lambda}_{i_1} = \tilde{\lambda}_{i_2} = 0$ if either $\tilde{g}^{i_1}$ or g^{i_2} is not binding, since in that case $\tilde{g}^{i_1\prime}(\bar{u}) = \tilde{g}^{i_2\prime}(\bar{u}) = 0$ by (87).

Conversely, suppose there exists a $\tilde{\lambda} \geqq 0$, with $\tilde{\lambda}_i = 0$ if $\tilde{g}(h)^i$ is not binding, and satisfying (84). We must show that there exist $\lambda \geqq 0$, with $\lambda_i = 0$

if g^i is not binding, and μ satisfying (83). We begin by assigning $\lambda_i = \tilde{\lambda}_i$ for each $g^i = \tilde{g}(h)^i$. Now g^i is binding if and only if the corresponding $\tilde{g}(h) = g^i$ is binding, so our hypothesis for this converse in part (b) implies that $\lambda_i = 0$ if g^i is not binding.

Next, for elements of $\tilde{g}(h)$ that are of the form $\tilde{g}^{i_1}$ or $\tilde{g}^{i_2}$ associated with some h^i, there are two cases to consider in defining a μ_i to satisfy (83).

If both $\tilde{g}^{i_1} = h^i$ and $\tilde{g}^{i_2} = -h^i$ are binding, then we define $\mu_i = \tilde{\lambda}_{i_1} - \tilde{\lambda}_{i_2}$. If either of $\tilde{g}^{i_1}$ or $\tilde{g}^{i_2}$ is not binding, then by (87) we have $\tilde{g}^{i_1\prime}(\bar{u}) = \tilde{g}^{i_2\prime}(\bar{u}) = h'(\bar{u}) = 0$, and we define $\mu_i = 0$.

With those λ_i and μ_i, (83) is satisfied.

□

Because of the Conversion Lemma, we can restrict discussion of path conditions to inequality constraints. The results will apply as well to equality and mixed constraints.

Proof of Theorem 3.

Parts (A) and (C) follow immediately from part (B) and the Conversion Lemma, by converting equality constraints $h = 0$ into inequality constraints $\tilde{g}(h) \geqq 0$, as in (82).

In the mixed case (A), for example, if the Mixed Tangency-Path Criterion holds for constraints (g, h), then, by the "only if" in (a.i) of the Conversion Lemma, the Inequalities Tangency-Path Criterion holds for the associated inequality constraints $\tilde{g}(h)$. So by the sufficiency proof below for part (B), for any f with partial derivatives there exist $\tilde{\lambda} \geqq 0$, with $\tilde{\lambda}_i = 0$ if $\tilde{g}(h)^i$ is not binding, satisfying (84); hence by the "if" in (b) of the Conversion Lemma there exist $\lambda \geqq 0$, with $\lambda_i = 0$ if g^i is not binding, and μ satisfying (83). A similar argument shows that the Equalities Tangency-Path Criterion is sufficient for Lagrange regularity with just equality constraints.

(B. Inequalities) For each j with $j \in J$, define:

$$\lambda_j = 0. \tag{89}$$

Then (80b) is satisfied. If the index set[53] I of binding constraints is empty, we are done, since then $\bar{u}$ is in the interior of the constraint set, so:

$$\frac{\partial f}{\partial u_j}(\bar{u}) = 0 \qquad (j = 1, \ldots, n), \tag{90}$$

which together with (89) satisfies (79). So we assume $I = \{1, \ldots, p\} \neq \emptyset$.

By the Reduction principle (16), it suffices to find a $\lambda \in I\!R^p$ satisfying $\lambda^T g_I'(\bar{u}) = -f'(\bar{u})^T$ and $\lambda \geqq 0$. Arguing by contradiction, suppose there

[53] I is defined in (7).

does not exist such a λ. Then the Fundamental Lemma implies there exists a $\tilde{z} \in I\!R^n$ such that:

$$g'_I(\bar{u})\tilde{z} \geqq 0 \tag{91a}$$
$$f'(\bar{u}) \cdot \tilde{z} > 0. \tag{91b}$$

By the Tangency-Path Criterion,[54] property (91a) implies the existence of a sequence of $w^i \in \operatorname{ch}(T_{\bar{u}}C(g))$ with:

$$w^i \xrightarrow[i\to\infty]{} \tilde{z}. \tag{92}$$

For all large enough i, therefore, (91b) implies:

$$f'(\bar{u}) \cdot w^i > 0. \tag{93}$$

By (138a), for each i there exist nonnegative $t_{i,1}, \ldots, t_{i,q}$ summing to 1, and there exist $w^{i,1}, \ldots, w^{i,q} \in T_{\bar{u}}C(g)$ with:[55]

$$w^i = t_{i,1}w^{i,1} + \cdots + t_{i,q}w^{i,q}. \tag{94}$$

It follows from (93) and (94) that:

$$f'(\bar{u}) \cdot w^{i,j} > 0 \tag{95}$$

for some large i and some $j = 1, \ldots, n$. Since $w^{i,j} \in T_{\bar{u}}C(g)$ and $w^{i,j} \neq 0$ (by (95)), it follows by (141) and the Proposition on Paths and Tangent Cones (146) that there is a path $\phi\colon T \subseteqq [0,1) \to C(g)$ with $\phi(0) = \bar{u}$ and $\phi'(0) = w^{i,j}$. Then:

$$\begin{aligned} \frac{d}{dt}|_{t=0}\, f(\phi(t)) &= f(\bar{u}) \cdot \phi'(0) \\ &= f(\bar{u}) \cdot w^{i,j} > 0. \end{aligned} \tag{96}$$

So, for some small $t > 0$,

$$f(\phi(t)) > f(\phi(0)) = f(\bar{u}), \tag{97}$$

contradicting that $\bar{u}$ maximizes f. □

Theorem 3 shows that the Tangency-Path Criterion[56] is strong enough for Lagrange regularity. Now we show it is not too strong.

---

[54] Page 124.

[55] Actually, Carathéodory's Theorem ensures that we can choose $q = n + 1$ (cf. [37], p. 155 (Theorem 17.1)). In fact, since $T_{\bar{u}}C(g)$ is a cone, we can choose $q = n$ ([37], p. 156, (Theorem 17.1.2)).

[56] See definition page 124.

Theorem 4. The Tangency-path criterion is necessary for Lagrange regularity. *Let $U \subseteqq I\!R^n$, and let $\bar{u} \in U$ be a limit point. Suppose $g\colon U \to I\!R^m$ and $h\colon U \to I\!R^k$ have partial derivatives at $\bar{u}$.*

A) (Mixed: Equalities and inequalities) *If (g,h) is Lagrange mixed-regular for $(\bar{u}, \mathcal{F}_D(\bar{u}))$, then (g,h) satisfies the Mixed Problem Tangency-Path Criterion.*

B) (Inequalities) *If g is Lagrange inequality-regular for $(\bar{u}, \mathcal{F}_D(\bar{u}))$, then g satisfies the Inequality-Problem Tangency-Path Criterion.*

C) (Equalities) *If h is Lagrange equality-regular for $(\bar{u}, \mathcal{F}_D(\bar{u}))$, then h satisfies the Equality-Problem Tangency-Path Criterion.*

Remark 6.3. Under the other hypotheses of Theorems 1 and 1^+, the Jacobian Criterion is strictly stronger than the Tangency-Path Criterion. There are two ways to see that the Jacobian hypotheses imply the Path Criterion. First, under the other hypotheses together with the Jacobian conditions, Theorems 1 and 1^+ imply Lagrange regularity, so Theorem 4 implies that the Path Criterion holds. More directly, from examining the proofs of Theorems 1 and 1^+ it is clear that the use we made of the Jacobian Criterion was precisely to obtain a path whose derivative satisfies the Path Criterion.

On the other hand, even with those other hypotheses, the Path Criterion does not imply the Jacobian Criterion. That is clear from Example 6.1, page 124, where the Jacobian Criterion clearly fails at $(0,0)$, but we can easily verify directly that the Path Criterion holds at $(0,0)$.[57] Alternatively, we can check that Lagrange regularity holds there,[58] so again by Theorem 4 the Path Criterion holds there.

Proof.

Parts (A) and (C) again follow from part (B) and the Conversion Lemma, by converting equality constraints $h = 0$ into inequality constraints $\tilde{g}(h) \geqq 0$, as in (82).

In the mixed case (A), for example, if (g,h) is Lagrange-regular, we convert the equality h constraint functions to obtain inequalities $\tilde{g}(h)$, including the original inequality constraints g. We then use the proof below for part (B)

---

[57] The vectors $v = (v_1, v_2)$ satisfying the hypothesis of the Path Criterion (72) are $v = (v_1, 0)$ for arbitrary v_1. On the other hand, the constraint set $C(g^1, g^2)$ is the part of the plane above the graph of the function $\psi(u_1)$ with values $u_1^2 \sin(1/u_1)$ for nonzero u_1, and with value 0 for $u_1 = 0$; so the paths from $(0,0)$ into the graph of the function ψ have derivative $(w_1, 0)$ for some w_1, and for every w_1 there is a path into the graph with that derivative. Thus any v satisfying the left hand side of (72) belongs to the right hand side.

[58] It is clear from the preceding footnote that any function f maximized at $(0,0)$ has $\partial f/\partial u_1 = 0$ there. Since we also have each $\partial g^i/\partial u_1 = 0$ as well as $\partial g^1/u_2 = 1$ and $\partial g^2/u_2 = -1$ there, it is easy to choose nonnegative λ_1 and λ_2 to obtain the Lagrange condition (11) for each such maximand f.

to establish that the Inequalities Tangency-Path Criterion (72) holds for $\tilde{g}(h)$. By the "if" part in (a.i) of the Conversion Lemma, it follows that the Mixed Tangency-Path Criterion (69) holds for (g, h). A similar argument shows that the Equalities Tangency-Path Criterion (73) is necessary for Lagrange regularity with just equality constraints.

(B. Inequalities) Suppose g is Lagrange inequality-regular for $(\bar{u}, \mathcal{F}_D(\bar{u}))$,[59] so the g^i have partial derivatives at $\bar{u}$. To verify (72), we must show: for any $z \in I\!R^n$,[60]

$$g_I'(\bar{u})z \geqq 0 \quad \Rightarrow \quad z \in V(g). \tag{98}$$

Without loss of generality, we will suppose that $\bar{u}$ is the origin.

a) If there are no binding constraints g^i, then $L(g) = \emptyset$, hence (98) is vacuously true. So now we suppose that there are binding constraints, i.e.,[61] the set $I = \{1, \ldots, p\}$ is nonempty.

b) For a proof by contradiction, suppose (98) is not true, so there is a $z \in L(g) \smallsetminus V(g)$. Now $V(g) = \mathrm{cl}(\mathrm{ch}(T_{\bar{u}}C(g)))$ is a closed convex cone, not equal to $I\!R^n$ (it does not contain z); so there exists[62] a $q \in I\!R^n$ such that:

$$q \neq 0 \tag{99a}$$

$$q \cdot u < 0 \qquad \text{for all } u \in V(g) \text{ with } -u \notin V(g) \tag{99b}$$

$$q \cdot u = 0 \qquad \text{for all } u \in V(g) \text{ with } -u \in V(g) \tag{99c}$$

$$q \cdot z > 0 \tag{99d}$$

$$q \cdot q = 1. \tag{99e}$$

We define the hyperplane H_q by:

$$H_q = \{u \in I\!R^n : q \cdot u = 0\}, \tag{100}$$

so q is orthogonal to H_q. And we define the subspace orthogonal to H_q:

$$Q = \{tq : t \in I\!R\}, \tag{101}$$

so $I\!R^n$ is the direct sum:

$$I\!R^n = H_q \oplus Q. \tag{102}$$

c) In what follows we shall define a differentiable function $f : I\!R^n \to I\!R$ that is maximized on $C(g)$ at $\bar{u} = 0$, and for which $f'(0) = q$. Then by part (a) above and Lagrange inequality-regularity, there must exist $\lambda_i \geqq 0$ such that $f'(0) = -\sum_{i=1}^{p} \lambda_i g^{i\prime}(0)$, so:

[59] See definition page 103.

[60] I is defined in (7).

[61] See the definition of binding constraint, page 101, and the notation established in (7) and (8).

[62] Cf. [26], p. 315, Theorem 2.7.

$$q \cdot z = f'(0) \cdot z \tag{103a}$$

$$= -\sum_{i=1}^{p} \lambda_i g^{i\prime}(0) \cdot z \tag{103b}$$

$$\leqq 0 \qquad \text{(by (70), since } z \in L(g)), \tag{103c}$$

which contradicts (99d).

d) Finding a function f as in (c) is equivalent to finding a function $f\colon I\!R^n \to I\!R$ that is differentiable at 0, has $f_u(0) = q$, and is maximized on $C(g)$ at 0.

e) Our intuition in defining f is simple. If all of $C(g)$ lies "below" the hyperplane H_q, then $f(x+tq) = t$ (for all $x \in H_q$ and $tq \in Q$) clearly satisfies the requirements of part (c) above. Now properties (99b,c) almost imply that $V(g)$, hence its subset $C(g)$, does lie below the hyperplane H_q. However they in fact allow $C(g)$ itself to rise "above" H_q — though only "gradually," much as the function $y = x^2$ rises above the x-axis. So we will define the graph of a function $\tilde{\gamma}$ below as the "upper boundary" of $C(g)$, and show it has (like the function x^2) a zero derivative at 0; then we define the level sets of the desired function f through shifts of the graph of $\tilde{\gamma}$.

f) We define:

$$\begin{aligned} B^k &= \{u \in U \colon \|u\| \leqq \frac{1}{2^k}\} \\ C^k &= C(g) \cap B^k \\ H_q^k &= \{u \in H_q \colon \|u\| \leqq \frac{1}{2^k}\}. \end{aligned} \tag{104}$$

g) To define the function f we will first define, for each $k \in I\!N$, a function $\gamma^k \colon H_q \to I\!R \cup \{-\infty\}$; then we will change γ^k into $\tilde{\gamma}^k$ so that its values belong to the ball B^k; finally, we will define f in terms of $\tilde{\gamma}^k$.

We define for each $k \in I\!N$, the function $\gamma^k \colon H_q \to I\!R \cup \{-\infty\}$ by:[63]

$$\gamma^k(x) = \begin{cases} \sup\{t \colon x + tq \in C^k\}, & \text{if } \exists t[x + tq \in C^k] \\ -\infty, & \text{otherwise.} \end{cases} \tag{105}$$

Because $C^k \subseteqq B^k$, the values of γ^k are either finite or $-\infty$. We note that:

$$\gamma^k(x) \geqq t \qquad \text{for all } x + tq \in C^k. \tag{106}$$

To obtain a function bounded below, we define:

[63] We use the supremum rather than the maximum because the sets C^k are not necessarily closed. That is because the constraint functions g^i are only assumed to have partial derivatives, and only at the origin, so they are not necessarily continuous in any neighborhood of 0.

$$\tilde{\gamma}^k(x) = \max\{0, \gamma^k(x)\} \qquad \text{for all } x \in H_q, \tag{107}$$

so:

$$\tilde{\gamma}^k \geqq \gamma^k \qquad \text{and} \qquad \tilde{\gamma}^k \geqq 0. \tag{108}$$

h) Next we show that for all large enough $k \in I\!N$:

$$\tilde{\gamma}^k(0) = 0. \tag{109}$$

First we note that for large enough $k \in I\!N$:

$$\neg \exists t_{\ 0<t\in I\!R}\ tq \in \mathrm{cl}(C^k). \tag{110}$$

Otherwise there would exist a sequence of points $(x^k + t^k q) \in C^k$ with $t^k > 0$ and:

$$\frac{\|x^k\|}{\|t^k q\|} \leqq \frac{1}{2^k} \xrightarrow[k\to\infty]{} 0 \tag{111}$$

and $x^k + t^k q \xrightarrow[k\to\infty]{} 0$ (since $x^k + t^k q \in C^k$). So by (140a) and compactness of the unit ball, there is a subsequence (re-using index k for convenience) with:

$$\frac{x^k + t^k q}{\|x^k + t^k q\|} \xrightarrow[k\to\infty]{} \text{some } \bar{w} \in T_0 C(g). \tag{112}$$

Since $\bar{w} \in T_0 C(g) \subseteqq V(g)$, properties (99b,c) imply:

$$q \cdot \bar{w} \leqq 0. \tag{113}$$

But using (111) we see that, in the maximum norm we have, for all large k:

$$\frac{x^k + t^k q}{\|x^k + t^k q\|} = \frac{x^k + t^k q}{\|t^k q\|} \xrightarrow[k\to\infty]{} \frac{q}{\|q\|}, \tag{114}$$

so $\bar{w} = q/\|q\|$. Then (113) implies $q \cdot q \leqq 0$, so $q = 0$, contradicting (99a) and completing the proof of (110). So for all large enough $k \in I\!N$ we have:

$$tq \in \mathrm{cl}(C^k) \ \Rightarrow \ t \leqq 0, \tag{115}$$

and therefore (109) holds.

i) Now we show that, for all sufficiently large $k \in I\!N$:

$$\tilde{\gamma}^k \text{ is differentiable at } 0 \in H_q \qquad \text{and} \qquad \tilde{\gamma}^{k\prime}(0) = 0. \tag{116}$$

Let $x^i \in H_q$ with $0 \neq \|x^i\| \xrightarrow[i\to\infty]{} 0$. We must show that:

$$\frac{|\tilde{\gamma}^k(x^i) - \tilde{\gamma}^k(0)|}{\|x^i - 0\|} \xrightarrow[i\to\infty]{} 0 \in I\!R, \tag{117}$$

which by (108) and (109) means:

$$\frac{\tilde{\gamma}^k(x^i)}{\|x^i\|} \xrightarrow[i\to\infty]{} 0. \tag{118}$$

We only need consider subsequences with $\tilde{\gamma}^k(x^i) > 0$, hence $\tilde{\gamma}^k(x^i) = \gamma^k(x^i)$, and show that any such "positive" subsequence itself has a subsequence satisfying (118). Now on any subsequence of a positive subsequence (indexed by "j"):

$$0 < \tilde{\gamma}^k(x^j) = \gamma^k(x^j) \xrightarrow[j\to\infty]{} 0, \tag{119}$$

since otherwise this subsequence itself would have a subsequence converging to some $t > 0$; but then (by (105)) there would exist $x^j + t^j q \in C^k$ converging to $tq \in \mathrm{cl}(C^k)$, contradicting (110).

Now (119) implies:

$$0 \neq \|x^j + \gamma^k(x^j)q\| \xrightarrow[j\to\infty]{} 0, \tag{120}$$

so by compactness of the unit ball, there is a subsequence of the j's on which:

$$\frac{x^j + \gamma^k(x^j)q}{\|x^j + \gamma^k(x^j)q\|} \xrightarrow[j\to\infty]{} \text{some } \bar{\bar{w}}. \tag{121}$$

We will show that (118) holds on any such subsequence (121). First note that:

$$\bar{\bar{w}} \in T_0C(g). \tag{122}$$

By (120) and (121), that would be immediate from (140) if $x^j + \gamma^k(x^j)q \in C^k$; but (105) only requires $\gamma^k(x^j)$ to be the sup of t for which $x^j + tq \in C^k$.[64] However, (105) and (121) imply there are t^j with $0 \neq x^j + t^j q \in C^k$, $\|x^j + t^j q\| \to 0$, and:

$$\frac{x^j + t^j q}{\|x^j + t^j q\|} \xrightarrow[j\to\infty]{} \bar{\bar{w}}, \tag{123}$$

so (122) holds.

From (122) and (99b,c) we see that:

$$0 \geqq q \cdot \bar{\bar{w}}. \tag{124}$$

Then (121) and (124) imply:

---

[64] See footnote 63, page 134.

$$\begin{aligned} 0 \geqq q \cdot \bar{\bar{w}} \xleftarrow[j \to \infty]{} q \cdot \frac{x^j + \gamma^k(x^j)q}{\|x^j + \gamma^k(x^j)q\|} \\ &= \frac{\gamma^k(x^j)}{\|x^j + \gamma^k(x^j)q\|} q \cdot q \qquad \text{(since } x^j \in H_q) \\ &= \frac{\gamma^k(x^j)}{\|x^j\| + \|\gamma^k(x^j)q\|} \qquad \text{(using (99e) and the sum norm)} \\ &\geqq 0 \qquad \text{(since } \gamma^k(x^j) > 0), \end{aligned} \tag{125}$$

which implies (118).

j) Now we define $f\colon \mathbb{R}^n \to \mathbb{R}$ by:[65]

$$f(u) = t - \tilde{\gamma}^K(x) \quad \text{for all } u \in \mathbb{R}^n \text{ with } u = x + tq \in H_q \oplus Q, \tag{126}$$

where K is any k so large that (109) and (116) hold, as in parts (h) and (i). Clearly $0 \in \mathbb{R}^n$ maximizes f on C^K, since $f(0) = 0$ by (126) and (109), and if $x + tq \in C^K \subseteqq C(g)$ then $f(x + tq) \leqq 0$ by (126), (106), and (108).

By parts (c) and (d) above, it only remains to show that f is differentiable at the origin $0 \in \mathbb{R}^n$, with $f_u(0) = q$. And for that it suffices to show:

$$\begin{aligned} f(x+tq) - f(0) &= f_u(0) \cdot (x + tq) + o(\|x + tq\|) \\ &= q \cdot (x + tq) + o(\|x + tq\|). \end{aligned} \tag{127}$$

Since $f(0) = 0$ and $q \cdot x = 0$ for $x \in H_q$, this becomes:

$$f(x + tq) = t + o(\|x + tq\|), \tag{128}$$

i.e.,

$$\frac{\gamma^K(x)}{\|x + tq\|} \xrightarrow[\|x+tq\| \to 0]{} 0. \tag{129}$$

Using the sum norm,[66] this becomes:

$$\frac{\gamma^K(x)}{\|x\| + \|tq\|} \xrightarrow[\|x+tq\| \to 0]{} 0, \tag{130}$$

which follows immediately from (116) and the fact that $\|x + tq\| \to 0$ implies both $\|x\| \to 0$ and $\|tq\| \to 0$, since $x \in H_q$, $tq \in Q$, and $H \perp Q$. □

---

[65] If we wanted f to have a unique maximum at the origin, we could instead define it by:

$$f(u) = t - \tilde{\gamma}^K(x) - \|x\|^2,$$

where $\|\cdot\|$ is the Euclidean norm.

[66] See (135c), page 139.

7. Appendix

Notation and terminology.

The set of natural numbers is $I\!N = \{0,1,2,3,\dots\}$, and the set of real numbers is $I\!R$. For $u \in I\!R^n = I\!R^r \times I\!R^s$ we write $u = (x, y)$, where $x \in I\!R^r$ and $y \in I\!R^s$. If $U \subseteqq I\!R^n$, then we say that $u \in I\!R^n$ is a *limit point* (or *accumulation point*) of U if every neighborhood of u contains a point of U other than u.

Suppose $F\colon U \subseteqq I\!R^n \to I\!R^p$ and $\bar{u} \in U$ is a limit point of U (which is not necessarily open). Then we say that F has a Fréchet derivative of F at $\bar{u}$ if: there exists a linear transformation $L\colon I\!R^n \to I\!R^p$ such that

$$F(u) = F(\bar{u}) + L(u - \bar{u}) + o(\|u - \bar{u}\|) \qquad \text{for all } u \in U. \tag{131}$$

We will denote such a linear transformation by $F_u(\bar{u})$. Differentiability will always mean Fréchet differentiability unless otherwise noted. When U is not an open set, the transformation L may not be unique.

Partial derivatives with respect to subspaces are defined for functions F at $\bar{u}$ in the usual way. When they exist, they may be denoted by $\dfrac{\partial F}{\partial u^i}(\bar{u})$, $F_x(\bar{x}, \bar{y})$, or $F_y(\bar{x}, \bar{y})$.

When partial derivatives $\partial F^i/\partial u_j(\bar{u})$ exist at $\bar{u}$, then $F'(u)$ denotes the $p \times n$ Jacobian matrix:

$$\begin{bmatrix} \dfrac{\partial F^1(u)}{\partial u_1} & \cdots & \dfrac{\partial F^1(u)}{\partial u_n} \\ \vdots & & \vdots \\ \dfrac{\partial F^p(u)}{\partial u_1} & \cdots & \dfrac{\partial F^p(u)}{\partial u_n} \end{bmatrix} \tag{132}$$

When a Fréchet derivative $F_u(\bar{u})$ exists, as a linear transformation it is represented, with respect to the standard bases of $I\!R^n$ by the matrix $F'(\bar{u})$, and so, for any $z \in I\!R^n$:

When F is Fréchet differentiable at $\bar{u}$, then $F'(\bar{u})z$ is a matrix representation of the vector $F_u(\bar{u})z$. (133)

We will often split the space $I\!R^n$ into two factors, $I\!R^n = I\!R^r \times I\!R^s$, and consider $X \subseteqq I\!R^r$ and $Y \subseteqq I\!R^s$ with $(\bar{x}, \bar{y}) \in X \times Y$. Then when $F\colon X \times Y \to I\!R^q$ and $\bar{x}$ is a limit point of X, if F has partial derivatives at $(\bar{x}, \bar{y})$ with respect to the X variables, then the corresponding $q \times r$ Jacobian matrix is denoted by $F'_x(\bar{x}, \bar{y})$; if $F(\,\cdot\,, \bar{y})$ has a Fréchet derivative at $\bar{x}$, it is denoted by $F_x(\bar{x}, \bar{y})$.

Analogously, when $\bar{y}$ is a limit point of Y and F has partial derivatives at $(\bar{x}, \bar{y})$ with respect to the Y variables, then the corresponding $q \times s$ Jacobian matrix is denoted by $F'_y(\bar{x}, \bar{y})$; if $F(\bar{x}, \cdot)$ has a Fréchet derivative at $\bar{x}$, it (or any representative) is denoted by $F_y(\bar{x}, \bar{y})$.

If A and B are linear subspaces of $I\!R^n$, and if every element u of $I\!R^n$ can be written uniquely as a sum of elements in A and B: $u = x + y$, where $x \in A$ and $y \in B$, then we say that $I\!R^n$ is the *direct sum* of A and B, and we write:

$$I\!R^n = A \oplus B. \tag{134}$$

All norms in a finite dimensional linear vector space lead to the same notions of convergence and differentiability; as convenient, we may use: for any $v \in I\!R^l$,

the *Euclidean norm*: $\|v\| = \sqrt{v_1^2 + \cdots + v_l^2}$ (135a)

the *maximum norm*: $\|v\| = \max\{|v_1|, \ldots, |v_l|\}$ (135b)

the *sum norm*: for any normed subspaces A and B, (135c)

if $v = a + b \in A \oplus B \ \& \ a \in A \ \& \ b \in B$,

then $\|v\| = \|a\| + \|b\|$.

For any $v \in I\!R^l$ and any real γ, we denote the closed γ-ball in $I\!R^l$ about v by:

$$B_\xi^{I\!R^l}(v) = \{v + w \in I\!R^l : \|w\| \leqq \xi\}, \tag{136}$$

or simply $B_\xi(v)$.

For x and y in $I\!R^l$:

$$\begin{aligned} x &\geqq y && \text{means } x_i \geqq y_i \text{ for all } i = 1, \ldots, n \\ x &\geq y && \text{means } x \geqq y \text{ and } x \neq y \\ x &> y && \text{means } x_i > y_i \text{ for all } i = 1, \ldots, n. \end{aligned} \tag{137}$$

For an open set $U \subseteqq I\!R^n$ and for $f = (f^1, \ldots, f^q) \colon U \to I\!R^q$, the notation $f \geqq 0$ means $f^i(x_1, \ldots, x_n) \geqq 0$ for each $i = 1, \ldots, q$ and for each $i = 1, \ldots, q$.

For any subset S of $I\!R^l$,

$$\begin{aligned}
\mathrm{ch}(S) &= \text{the convex hull of } S && \text{(138a)}\\
&= \{t_0x_0 + t_1x_1 + \cdots + t_mx_m \colon m \in I\!N \;\&\; x_0, x_1, \ldots, x_m \in S\\
&\qquad \&\; t_0, t_1, \ldots, t_m \geqq 0 \;\&\; t_0 + t_1 + \cdots + t_m = 1\}\\
\mathrm{cl}(S) &= \text{the topological closure of } S && \text{(138b)}\\
\mathrm{cone}(S) &= \text{the conical closure of } S && \text{(138c)}\\
&= \{tx \colon x \in S \;\&\; \text{real } t \geqq 0\}\\
\mathrm{wedge}(S) &= \mathrm{cone}(\mathrm{ch}(S)) && \text{(138d)}\\
&= \text{the convex cone generated by } S\\
\mathrm{span}(S) &= \text{the linear subspace generated by } S. && \text{(138e)}
\end{aligned}$$

For any subsets S and T of $I\!R^l$:

$$\begin{aligned}
S + T &= \{x + y \colon x \in S \;\&\; y \in T\}\\
S + \emptyset &= S = \emptyset + S.
\end{aligned} \qquad (139)$$

Elements $x \in I\!R^l$ will be treated as column vectors for matrix multiplication; then we denote the transpose by x^T, which we treat as a row vector.

The tangent cone. To describe derivatives of paths, we use the tangent cone. For any set $S \subseteqq I\!R^n$ with $\bar{u} \in S$, a vector $v \in I\!R^n$ is *tangent to* S *at* $\bar{u}$ if:
either:

a) there exists a sequence of points $u^i \in S$ such that: (140a)
  i) $u^i \neq \bar{u}$ for all $i = 1, 2, 3,$
  ii) $\|u^i - \bar{u}\| \xrightarrow[i\to\infty]{} 0$
  iii) $\frac{u^i - \bar{u}}{\|u^i - \bar{u}\|} \xrightarrow[i\to\infty]{} \frac{v}{\|v\|}$;

or:

b) $\bar{u}$ is an isolated point of S and $v = 0$. (140b)

We define:

$$T_{\bar{u}}S = \{rv \in I\!R^n \colon v \text{ is tangent to } S \text{ at } \bar{u} \;\&\; r \text{ is a nonnegative real number}\}. \qquad (141)$$

As this is clearly a cone, $T_{\bar{u}}S$ is called the *tangent cone of* S *at* $\bar{u}$. (For an interpretation by paths, see the Proposition on Paths and Tangent Cones below.)

Algebra. Our understanding of constraint qualifications, and our proofs of Lagrange regularity rest on an algebraic theorem of the alternative.

Theorem 5. Theorem of the alternative. *Let A, B, and C be matrices whose components are from* $I\!R$. *Suppose that*

A has a rows and n columns, and $\alpha \in I\!R^a$,
B has b rows and n columns, and $\beta \in I\!R^b$,
C has c rows and n columns, and $\gamma \in I\!R^c$,

where $0 < a \in I\!N$, $0 < b \in I\!N$, $0 < c \in I\!N$, *and* $0 < n \in I\!N$.
Then:

I) Exactly one of (1) or (2) is true:

1) There exists $r = (r_1, ..., r_n) \in I\!R^n$ *solving:*

$$Ar > \alpha$$
$$Br \geqq \beta$$
$$Cr = \gamma.$$

2) There exists $u \in I\!R^a$, $v \in I\!R^b$, *and* $z \in I\!R^c$, *such that both (a) and (b) hold:*

a) $u^T A + v^T B + z^T C = 0$

b) either (i) or (ii) hold:

i) $u \geq 0 \;\&\; v \geqq 0 \;\&\; u^T\alpha + v^T\beta + z^T\gamma \geqq 0$

or

ii) $u = 0 \;\&\; v \geqq 0 \;\&\; v^T\beta + z^T\gamma > 0.$

II) When some, but not all, of (A, α) *or* (B, β) *or* (C, γ) *are not present, then the same alternatives (1) and (2) in part (I) above hold with these modifications:*

In (I,1):

remove the first row if A *and* α *are not present;*
remove the second row if B *and* β *are not present;*
remove the first row if C *and* γ *are not present.*

In (2,a):

set $A = 0$ *and* $u = 0$ *if* A *and* α *are not present;*
set $B = 0$ *and* $v = 0$ *if* B *and* β *are not present;*
set $C = 0$ *and* $z = 0$ *if* C *and* γ *are not present.*

In (2,b):

only case (ii) occurs if A *and* α *are not present;*
set $\beta = 0$ *and* $v = 0$ *if* B *and* β *are not present;*
set $\gamma = 0$ *and* $z = 0$ *if* C *and* γ *are not present.*

The theorem can be proved from the Transposition Theorem of Motzkin [35], [33], pp. 28(2)–29, which is a homogeneous version (equivalent to setting $\alpha = 0$, $\beta = 0$, and $\gamma = 0$). A proof by Fourier elimination, (cf. [39], pp. 1–20, [28]) shows that it applies to arbitrary ordered fields. Cf. also [34], p. 181.

Analysis. We use an implicit function theorem with weaker than standard differentiability hypotheses. This is a special case of Theorem 1 of [21].

Theorem 6. A Non-C^1 implicit function theorem. *Let X be a subset of $I\!R^m$, let Y be a subset of $I\!R^k$, and let $U \subseteqq X \times Y$. Let $\bar{u} = (\bar{x}, \bar{y}) \in U$, where $\bar{x} \in X$ is a limit point of X, and $\bar{y} \in Y$. Suppose that $\psi\colon X \times Y \to I\!R^k$ and:*

$$\psi(\bar{x}, \bar{y}) = 0; \tag{142a}$$

$$\psi(\,\cdot\,,\,\cdot\,) \text{ is differentiable at } (\bar{x}, \bar{y}), \tag{142b}$$
$$\text{with partial derivatives } \psi_x(\bar{x}, \bar{y}) \text{ and } \psi_y(\bar{x}, \bar{y});$$

$$\psi_y(\bar{x}, \bar{y}) \text{ is surjective; i.e.,} \tag{142c}$$

$$\det\left(\begin{bmatrix} \dfrac{\partial \psi^1(\bar{x}, \bar{y})}{\partial y_1} & \cdots & \dfrac{\partial \psi^1(\bar{x}, \bar{y})}{\partial y_k} \\ \vdots & & \vdots \\ \dfrac{\partial \psi^k(\bar{x}, \bar{y})}{\partial y_1} & \cdots & \dfrac{\partial \psi^k(\bar{x}, \bar{y})}{\partial y_k} \end{bmatrix}\right) \neq 0.$$

Suppose that, for some $\gamma > \|(\psi_y(\bar{x}, \bar{y}))^{-1}\psi_x(\bar{x}, \bar{y})\|$:[67]

$$x \in X \text{ and } y \in B^{I\!R^k}_{\gamma\|x\|}(\bar{y}) \quad \Rightarrow \tag{142d}$$
$$(x, y) \in U \text{ and } \psi(x,\,\cdot\,) \text{ is defined and continuous on } B^{I\!R^k}_{\gamma\|x\|}(\bar{y}).$$

For any $\xi > 0$ and $x \in X$, let $X_\xi = B^{I\!R^m}_\xi(\bar{x}) \cap X$.
Then:

a) *There exists a real $\bar{\delta} > 0$ and a function $\phi \in \prod_{x \in X_{\bar{\delta}}} B^{I\!R^k}_{\gamma\|x\|}(\bar{y})$ such that: for all $x \in X_{\bar{\delta}}$,*

$$\psi(x, \phi(x)) = 0 \tag{143a}$$
$$\phi(\bar{x}) = \bar{y}. \tag{143b}$$

b) *The $\bar{\delta}$ in part (a) can be chosen so that every function $\phi \in \prod_{x \in X_\xi} B^{I\!R^k}_{\gamma\|x\|}(\bar{y})$ satisfying (143) is differentiable at $\bar{x}$.*
c) *For every $\xi > 0$, all functions $\phi \in \prod_{x \in X_\xi} B^{I\!R^k}_{\gamma\|x\|}(\bar{y})$ satisfying (143) that are differentiable at $\bar{x}$ have the same derivative $\phi_x(\bar{x})$:*

$$\phi_x(\bar{x}) = -\,(f_y(\bar{x}, \bar{y}))^{-1} f_x(\bar{x}, \bar{y}). \tag{144}$$

A path interpretation of the tangent cone. The tangent cone of a set can be interpreted as the set of derivatives of paths into the set.

Let $\bar{u} \in S \subseteqq I\!R^n$, and suppose that $\bar{u}$ is a limit point of S. Suppose $0 \in T \subseteqq [0, 1]$ and 0 is a limit point of T and $\phi\colon T \to S$ has $\phi(0) = \bar{u}$; then we

[67] By the norm of a linear transformation L, we mean $\|L\| = \max\{\|Lx\|\colon \|x\| = 1\}$.

call (ϕ, T) a *path from* $\bar{u}$ *into* S. If also ϕ has a derivative $v = \phi'(0) \in I\!R^n$ at $0 \in T$, then we say that v is a *tangent of the path* (ϕ, T). We denote by $D_{\bar{u}}(S)$ the set of all tangents of paths from $\bar{u}$ into S:

$$D_{\bar{u}}(S) = \{v \in I\!R^n : \exists (\phi, T)_{\ (\phi,T) \text{ is a path from } \bar{u} \text{ into } S}\ v = \phi'(0)\}. \qquad (145)$$

Proposition on paths and tangent cones. If $\bar{u} \in S \subseteq I\!R^n$ and if $\bar{u}$ is a limit point of S, then:

$$T_{\bar{u}}(S) = D_{\bar{u}}(S). \qquad (146)$$

So every tangent vector to S at $\bar{u}$ is a tangent of a path from $\bar{u}$ into S, and vice versa. (Such paths need not be continuous.)

8. Historical comments and comparisons

The beginnings. Lagrange illustrated for one equality constraint, and suggested for multiple constraints a "general principle":[68]

> When a function of several variables is to have a maximum or a minimum, and when there are one or more equations among these variables, it will suffice to add to the proposed function the functions which must vanish, each multiplied by an undetermined quantity, and then to seek the maximum or minimum as if the variables were independent; the equations that one will find, combined with the given equations, will serve to determine all the unknowns.

Euler [12] has been credited ([7], [9], [14], [27]) with originating a principle ("Euler's rule"), precursor of the Lagrange approach, for extremizing functionals subject to constraints. This was in the context of the calculus of variations.

We will sketch the history of constraint qualifications separately for Jacobian conditions and path conditions. To simplify the comparison, we focus primarily on inequality constraints.[69]

8.1 Jacobian constraint qualifications

We outline a sequence of increasingly weaker Jacobian constraint qualifications. For simplicity, the constraint qualifications should be interpreted as applying only to binding constraints.

---

[68] Our translation of the end of Section 58, in Chapter XI of the Second Part of [31]. An earlier exposition of the idea is contained in [10], Quatrieme Section, paragraphs 1-8, pages 44–49.

[69] We only discuss Lagrange regularity. See p. 103 for brief notes on quasi-regularity.

Classical. To obtain Lagrange multipliers, modern treatments of equality-constrained optimization have assumed the constraint functions are C^1 and have a Jacobian of maximum rank; this allows application of the classical implicit function theorem, which uses a C^1 hypothesis. Cf. Bolza [7],[70] Carathéodory [9], Bliss [5], [6], as well as many recent textbooks.

1917. Hancock suggests handling constrained maximization problems with inequality constraints ("limitations") by converting them into equality problems [17], p. 150. A constraint $g(x) \geqq 0$ would be replaced by the equality $g(x) + z^2 = 0$.

1937. Valentine uses the same conversion device in treating a calculus of variations minimization problem with inequality side conditions [41]. He also determines the sign of the Lagrange coefficients for the inequality constraints.

1939. Karush's Master of Science dissertation [25] went unnoticed for many years, although it contains most of the basic concepts and many of the results of later work [25]. He uses the same squared slack variable device as Valentine had[71] to prove his Theorem 3:1 (pp. 11–13), which proved Lagrange quasi-regularity for the inequalities case. Following his Theorem 3:1, which proves Lagrange quasi-regularity for the inequalities case, he notes that with "normality" (the classical rank condition) and C^1 constraints, the Lagrange equations would follow; and if the constraints were C^2, the multipliers would be of the usual sign.[72]

Karush's Theorem 3:3 shows that the positivity condition $\exists\xi[g'(\bar{u})\xi > 0]$ of (24a) is sufficient for Lagrange inequality-regularity, assuming that all the constraints are C^1.

1953. Pennisi gives constraint qualifications for the mixed problem [36]. Assuming that all constraints are C^2, are effective, and are satisfied with equality at $\bar{u}$, he obtains Lagrange regularity for the mixed problem by assuming that the matrix $(g'(\bar{u}), h'(\bar{u}))$ has full rank (i.e., $m + k$).

1956. Arrow and Hurwicz obtain Lagrange mixed-regularity, using the maximum rank condition under C^1 assumptions [2].

1961. Arrow, Hurwicz, and Uzawa independently discover Karush's positivity condition $\exists\xi[g'(\bar{u})\xi > 0]$ (cf. (24a)). Their Theorem 3 shows it is sufficient for Lagrange inequality-regularity. (The centrality of this condition is witnessed not only by its independent rediscovery, but by the fact that, by our

---

[70] Bolza's formulation makes clear that C^1 is a sufficient smoothness condition for Lagrange regularity. Earlier contributions appear to postulate real-analyticity, as in Weierstrass [43].

[71] Neither Valentine nor Karush seem to refer to Hancock, and Karush does not refer to Valentine.

[72] He remarks (p. 13) that it is possible to reduce the C^2 assumption to C^1.

Theorems 1.B and 2.B, it falls short of minimality only in lacking the alternative (24b).) The smoothness assumptions on the constraints g were reduced to differentiability at $\bar{u}$.

1967. Mangasarian and Fromovitz consider the mixed problem [32]. The smoothness assumptions of Pennisi are reduced from C^2 to C^1, and Pennisi's full rank condition is reduced to the classical full rank condition on just $h'(\bar{u})$, together with the existence (as in (23a) above) of a vector ξ such that $g'(\bar{u})\xi > 0$ and $h'(\bar{u})\xi = 0$.

1969. Mangasarian [33], p. 173, Theorem 6, part (iv) also considers the mixed problem, obtaining Lagrange mixed-regularity while reducing the C^1 constraints on g to differentiability at $\bar{u}$, but maintaining the C^1 hypothesis on h. (In Theorem 1.A above, the C^1 smoothness is reduced still further.)

1974. Halkin [16], p. 235, section 3 weakens the C^1 smoothness conditions used by earlier writers for their constraint functions, and he allows the underlying space X to be a not-necessarily-finite-dimensional normed linear space. He obtains quasi-regularity results through a new proof of the implicit function theorem using differentiability, continuity, and Brouwer's fixed point theorem, rather than the usual C^1 hypothesis and the contraction mapping theorem. Unaware of his earlier result, we embarked on a similar quest, using very similar techniques [22]. The lighter continuity assumptions we present here in Theorem 1^+ (for finite dimensional spaces) however, provide some extensions beyond the "Multiplier Rule" in his Section 3.

While our Theorem 1 uses essentially the same smoothness hypotheses for constraint functions as his Multiplier Rule (the classes $\mathcal{G}_D$ and $\mathcal{H}_{DC}$, requiring continuity in a neighborhood), our Theorem 1^+ weakens even further the continuity hypotheses for the equality constraints (allowing discontinuities in every neighborhood) as shown in our Example 4.2.[73] (Our Example 4.3 shows that no weaker continuity condition of our type would suffice for Lagrange regularity.)

His multiplier conclusion is just Lagrange quasi-regularity, rather than the full regularity conclusion of our Theorems 1, 1^+, 2, 3, and 4. The weaker quasi-regularity conclusion does not require any constraint qualifications, but it is less informative than full regularity in two respects: it does not tell when the

[73] Halkin's ability to weaken continuity assumptions for equality constraints below the C^1 properties of earlier authors is allowed by his introduction of an implicit function theorem with weaker assumptions properties [16], Theorem D than earlier authors had used. Our ability to weaken them still further is allowed by our Non-C^1 Implicit Function Theorem (page 142), which though similar has even weaker smoothness assumptions. Cf. [21].

coefficient λ_0 in (13) is nonzero, and it does not say that the λ_i coefficients for the nonbinding g_i in (10b) are zero.[74]

Halkin allows the underlying space X to be an arbitrary normed linear space, rather than the finite dimensional spaces we postulate. Examination of the proof (in [21]) of our Non-C^1 Implicit Function Theorem, and the applications we make of it in our Theorems 1 and 1^+, suggest that a similar extension of all our theorems is possible.[75]

Finally, Halkin discusses only sufficiency of his conditions, rather than the minimality notions of our Theorem 2 and the necessity conditions of Theorem 4.

1995. Minimality. Summarizing, we have:

$$
\begin{array}{ll}
\operatorname{rank}(g'(\bar{u})) = m & \text{(Karush } (C^2)\text{, Pennisi } (C^2)\text{, A-H } (C^1)\text{, A-H-U } (D^1)) \\
\quad\Downarrow \;\not\Uparrow & \\
\exists\xi[g_I{}'(\bar{u})\xi > 0] & \text{(Karush } (C^1)\text{, A-H-U } (D^1)) \\
\quad\Downarrow \;\not\Uparrow & \\
\multicolumn{2}{l}{[\exists\xi[g_I{}'(\bar{u})\xi > 0]] \text{ or } [\operatorname{wedge}(a(1),\ldots,a(p)) = I\!R^n]} \\
 & \text{(Theorem 1.B above).}
\end{array}
\tag{147}
$$

How much further does the weakening chain go? Theorem 2 says the chain stops here, since the last property is a minimal sufficient condition. Parts (A) and (C) of Theorem 2 show that analogous minimality results hold for Lagrange mixed-regularity and equality-regularity.

8.2 Path constraint qualifications

We outline a sequence of increasingly weaker path constraint qualifications.

1939. Karush obtains Lagrange inequality regularity in his Theorem 3:2, assuming C^1 constraints and imposing a path condition ("Property Q") that we can write as: $L(g) \subseteqq A(C(g))_{C^1}$, denoting by $A(C(g))_{C^1}$ the set of derivatives (i.e. directions) of C^1 "arcs," or paths from $\bar{u}$ into the constraint set.

---

[74] Property $\lambda_0 \neq 0$ would follow under the special Jacobian rank condition (31) of Corollary 1a: otherwise the quasi-regularity condition (13) would violate the quasi-regularity condition $(\lambda_0, \lambda, \mu) \neq 0$.

[75] Theorem V.3.3.2 in Hurwicz [20] establishes a full Lagrange regularity ("quasi-saddle-point") for constraints in Banach spaces, using an infinite-dimensional extension of the Kuhn-Tucker constraint qualification.

Ioffe and Tihomirov [23] contains extensions of these and other results to infinite dimensional spaces, as well as references to other literature.

1951. Kuhn and Tucker's Theorem 1 independently retraces part of the path traveled earlier by Karush. They prove Lagrange inequality-regularity using the weaker assumption of differentiability for constraints, rather than C^1. The Kuhn-Tucker "constraint qualification" says $L(g) \subseteqq A(C(g))_{D^1}$, where $A(C(g))_{D^1}$ is the set of derivatives of paths from $\bar{u}$ into the constraint set that are differentiable at $\bar{u}$ [29].[76] They provide more information about the Lagrange multipliers (analogous to our condition (10b)). Kuhn later became aware of Karush's work and earlier references to it ([11], [36], [40], et al.), and related research by others. See [30] for his very informative historical account, and some general comments on the several quite divergent interests that converged in similar theorems.

1961. Arrow, Hurwicz, and Uzawa weaken the constraint qualification further, assuming only that $L(g) \subseteqq \mathrm{cl}(\mathrm{ch}(A(C(g))_{D^1}))$. Calling this "Constraint Qualification W," their Theorem 1 proves it is sufficient for Lagrange inequality-regularity.

1967. Abadie weakens the Kuhn-Tucker Constraint Qualification in a different way. He assumes that $L(g) = T_{\bar{u}}(C(g))$ and that all the constraints were differentiable at $\bar{u}$; and he proves Theorem 4 of [1], Lagrange mixed-regularity.[77]

1971. Gould and Tolle weaken the constraint qualifications of both Abadie and Arrow, Hurwicz, and Uzawa. They assume $A(C(g))_{D^1} = \mathrm{cl}(\mathrm{ch}(T_{\bar{u}}(C(g))))$ and prove Lagrange mixed-regularity, assuming both differentiability and local continuity of the constraints at $\bar{u}$. This applies the closure and convex hull operations used by Arrow, Hurwicz, and Uzawa, but to the tangent cone $T_{\bar{u}}(C(g))$ used by Abadie [1], rather than to the smaller set $A(C(g))_{D^1}$ of [4].

1995. The Tangency-path Path Criterion of page 124 above weakens the constraint qualification further. Requiring just existence of partial derivatives at $\bar{u}$, it requires only $L(g) \subseteqq \mathrm{cl}(\mathrm{ch}(T_{\bar{u}}(C(g))))$. The stronger equality property used by Gould and Tolle follows automatically with their differentiability assumptions (according to Abadie's Lemma 4), but not with our weaker partial differentiability assumptions, as can be seen from the following example:[78]

Example 8.1.

---

[76] Their main results are formulated for the case of inequalities, where the variables x are also required to be nonnegative. Certain other types of constraints are briefly dealt with in their Section 8.

[77] Actually the proof of his Theorem 4 only uses $A(C(g))_{D^1} \subseteqq T_{\bar{u}}(C(g))$, but under his differentiability assumption his Lemma 4 shows that the inclusion automatically holds in the opposite direction.

[78] See also (75), p. 126, and (76), p. 126.

$$g^1(x_1, x_2) = \begin{cases} x_1, & \text{for } x_2 = 0 \\ 0, & \text{otherwise.} \end{cases}$$
$$g^2(x_1, x_2) = \begin{cases} x_2, & \text{for } x_1 = 0 \\ 0, & \text{otherwise.} \end{cases} \tag{148}$$

This pair (g^1, g^2) fails to satisfy the Fréchet differentiability and continuity conditions that Gould and Tolle impose, as well as their $L(g) = V(g)$ constraint qualification. Nevertheless, it is partially differentiable at the origin $(0, 0)$, and satisfies our criterion (72) since $L(g) \subsetneqq V(g)$,[79] and hence is Lagrange inequality-regular.

We developed our Tangency-Path Criterion[80] based in part on Hestenes' use of the tangent cone ([18], pp. 25 ff., [19], pp. 203 ff.) and in part by analogy with Constraint Qualification W of Arrow, Hurwicz, and Uzawa [4]. Subsequently we discovered the paper by Gould and Tolle [15] with its closely related, but stronger constraint qualification.

Despite the stronger hypotheses of Gould and Tolle, and their more demanding constraint qualification, the proofs we had developed for our Theorems 3 and 4 were similar to certain parts of Gould and Tolle's proof.

Summarizing, we have:

$$\begin{array}{ll}
L(g) \subseteqq A(C(g))_{C^1} & \text{(Karush } (C^1)) \\
\Downarrow \;\; \not\Uparrow & \\
L(g) \subseteqq A(C(g))_{D^1} & \text{(Kuhn-Tucker } (D^1)) \\
\Downarrow \;\; \not\Uparrow & \\
L(g) \subseteqq \begin{cases} \mathrm{cl}(\mathrm{ch}(A(C(g))_{D^1})) \\ T_{\bar{u}}(C(g)) \end{cases} & \begin{array}{l} \text{(A-H-U } (D^1)) \\ \text{(Abadie } (D^1)) \end{array} \\
\Downarrow \;\; \not\Uparrow & \\
L(g) = \mathrm{cl}(\mathrm{ch}(T_{\bar{u}}(C(g)))) & \text{(Gould-Tolle } (D^1)) \\
\Downarrow \;\; \not\Uparrow \quad \text{(Example 8.1)} & \\
L(g) \subseteqq \mathrm{cl}(\mathrm{ch}(T_{\bar{u}}(C(g)))) & \text{(our Theorem 3 (partial derivatives)).}
\end{array} \tag{149}$$

How much further does the weakening chain go? Theorem 4 says the chain stops here, since the last property is necessary for Lagrange inequality-regularity.

---

[79] Note that $g'(0, 0)$ is the identity matrix, so $L(g)$ is the nonnegative quadrant of $I\!R^2$. Since $C(g)$ is all of $I\!R^2$ except the negative rays of the axes, $V(g)$ includes much more than the nonnegative quadrant, so our Theorem 3 implies that g is Lagrange inequality-regular.

[80] Page 124.

Parts (A) and (C) of Theorem 4 show that analogous necessity results hold for Lagrange mixed-regularity and equality-regularity.

The tangent cone was applied to optimization problems by Hestenes [18], [19], Abadie [1], and Varaiya [42]. It was defined by Bouligand [8], paragraph 68, pp. 65–66 (as the *contingent* set).

References

1. Abadie, J.: On the Kuhn-Tucker Theorem. In: Nonlinear Programming, North-Holland, New York 1967
2. Arrow, K.J., Hurwicz, L.: Reduction of constrained maxima to saddle-point problems. In: Proceedings of the Third Berkeley Symposium on Mathematical Statistics and Probability. volume V, pp.1-20 University of California Press, Berkeley and Los Angeles 1956
3. Arrow, K.J., Hurwicz, L. eds.: Studies in Resource Allocation Processes. Cambridge University Press, Cambridge 1978
4. Arrow, K.J., Hurwicz, L., Uzawa, H.: Constraint qualifications in maximization problems. Naval Research Logistics Quarterly **8**, 175-191 (1961)
5. Bliss, G.A.: Normality and abnormality in the calculus of variations. Transactions of the American Mathematical Society **43**, 365-376 (1938)
6. Bliss, G.A.: Lectures on the Calculus of Variations. University of Chicago Press, Chicago 1946
7. Bolza, O.: Vorlesungen über Variationsrechnung. Chelsea Publishing Company, New York (second edition, no date, author's preface dated 1909)
8. Bouligand, G.: Introduction à la Géométrie Infinitésimale Directe. Librairie Vuibert, Paris 1932
9. Carathéodory, C.: Calculus of Variations and Partial Differential Equations of the First Order. Chelsea Publishing Company, New York 1982 (Second (revised) edition originally published as Variationsrechnung und Partielle Differentialgleichungen erster Ordnung, B.G.Teubner, Berlin 1935)
10. de la Grange, J.L.: Méchanique Analitique. Chez la Veuve Desaint, Libraire, Paris, 1788 (See also Lagrange, J.L.)
11. El-Hodiri, M.A.: The Karush characterization of constrained extrema of functions of a finite number of variables. Series A, No. 3, UAR Ministry of Treasury Research Memoranda, July 1967
12. Euler, L.: Methodus Inveniendi Lineas Curvas. In: Leonhardi Euleri Opera Omnia, series prima XXIV, vol. 34. Swiss Society of Natural Sciences, Bern 1952 (Reprinted from Methodus Inveniendi Lineas Curvas, Lausanne Geneva 1744)
13. Evans, J.P.: On constraint qualifications in nonlinear programming. Report 6917, Center for Mathematical Studies in Business and Economics, Graduate School of Business, University of Chicago, Chicago, May 1969
14. Goldstine, H.H.: A History of the Calculus of Variations. Springer-Verlag, New York 1980
15. Gould, F.J., Tolle, J.W.: A necessary and sufficient qualification for constrained optimization. SIAM Journal of Applied Mathematics **20**, 164-172 (1971)
16. Halkin, H.: Implicit functions and optimization problems without continuous differentiability of the data. SIAM Journal on Control **12**, 229-236 (1974)

17. Hancock, H.: Theory of Maxima and Minima. Ginn and Company, Boston 1917
18. Hestenes, M.R.: Calculus of Variations and Optimal Control Theory. John Wiley & Sons, New York 1966
19. Hestenes, M.R.: Optimization Theory. John Wiley & Sons, New York 1975
20. Hurwicz, L.: Programming in Linear Spaces. In: Studies in Linear and Non-Linear Programming (Kenneth J. Arrow, Leonid Hurwicz, Hirofumi Uzawa eds.). Chapter 4, pp.38-102 Stanford University Press, Stanford 1958
21. Hurwicz, L., Richter, M.K.: Implicit Functions and Diffeomorphisms without C^1. Discussion Paper No. 279, Department of Economics, University of Minnesota 1994 (see also this volume, pp.65-96)
22. Hurwicz, L., Richter, M.K.: Optimization and Lagrange Multipliers: Non-C^1 Constraints and "Minimal" Constraint Qualifications. Discussion Paper No. 280, Department of Economics, University of Minnesota 1995
23. Ioffe, A.D., Tihomirov, V.M.: Theory of Extremal Problems. North-Holland, Amsterdam 1979
24. John, F.: Extremum problems with inequalities as subsidiary conditions. In: Studies and Essays: Courant Anniversary Volume (K. O. Friedrichs, O. E. Neugebauer, J. J. Stoker eds.). pp.187-204 Interscience Publishers, New York 1948
25. Karush, W.: Minima of functions of several variables with inequalities as side conditions. Master's dissertation 1939
26. Klee, V.L. Jr.: Separation properties of convex cones. Proceedings of the American Mathematical Society **6**, 313-318 (1955)
27. Kneser, A.: Variationsrechnung. Teubner, Leipzig, 1899-1916 (Originally published as Heft 5 1904)
28. Kuhn, H.W.: Solvability and consistency for linear equations and inequalities. American Mathematical Monthly **63**, 217-232 (1956)
29. Kuhn, H.W., Tucker, A.W.: Nonlinear programming. In: Second Berkeley Symposium on Mathematical Statistics and Probability (J. Neyman ed.). pp.481-492 University of California Press 1951
30. Kuhn, H.W.: Nonlinear Programming: A Historical View. In: Nonlinear Programming (Richard W. Cottle, C. E. Lemke eds.). American Mathematical Society 1976
31. Lagrange, J.L.: Théorie des Fonctions Analytiques. Courcier, Paris 1813, Nouvelle édition. See also de la Grange, J.L.)
32. Mangasarian, O.L., Fromovitz, S.: The Fritz John necessary optimality conditions in the presence of equality and inequality constraints. Journal of Mathematical Analysis and Applications **17**, 37-47 (1967)
33. Mangasarian, O.L.: Nonlinear Programming. McGraw-Hill, New York 1969
34. McFadden, D., Richter, M.K.: Stochastic rationality and revealed stochastic preferences. In: Preferences, Uncertainty, and Optimality: Essays in Honor of Leonid Hurwicz (John S. Chipman, Daniel McFadden and Marcel K. Richter eds.). chapter 6. Westview Press, Boulder, CO 1990
35. Motzkin, T.S.: Beiträge zur Theorie der linearen Ungleichungen. Dissertation, Basle 1934
36. Pennisi, L.L.: An indirect sufficiency proof for the problem of Lagrange with differential inequalities as added side conditions. Transactions of the American Mathematical Society **74**, 177-198 (1953)
37. Rockafellar, R.T.: Convex Analysis. Princeton University Press, Princeton 1970
38. Slater, M.: Lagrange Multipliers Revisited: A Contribution to Nonlinear Programming. Cowles Commission Discussion Paper, Math. 403, November 1950
39. Stoer, J., Witzgall, C.: Convexity and Optimization in Finite Dimensions, volume I. Springer-Verlag, New York 1970

40. Takayama, A.: Mathematical Economics. The Dryden Press, Hinsdale, IL 1974
41. Valentine, F.A.: The problem of Lagrange with differential inequalities as added side conditions. In: Contributions to the Calculus of Variations 1933-1937: Theses Submitted to the Department of Mathematics of the University of Chicago, pp.403-447. University of Chicago Press, Chicago 1937
42. Varaiya P.P.: Nonlinear programming in Banach space. SIAM Journal of Applied Mathematics **15**, 284-293 (1967)
43. Weierstrass, K.: Mathematische Werke. Akademische Verlagsgesellschaft, Leipzig 1927 (Edited from notes of Weierstrass's 1875-1878 lectures.)

Adv. Math. Econ. 5, 153–166 (2003)

Advances in
MATHEMATICAL
ECONOMICS

Monte Carlo method for pricing of Bermuda type derivatives

Shigeo Kusuoka

Graduate School of Mathematical Sciences, The University of Tokyo, 3-8-1 Komada, Meguro-ku, Tokyo 153-8914, Japan

Received: August 2, 2002

JEL classification: C63, G12

Mathematics Subject Classification (2000): 65C05, 60G40

Abstract. The author gives a rigorous justification of a certain Monte Carlo Method for pricing of Bermuda type derivatives given by Longstaff-Schwartz.

Key words: Mathmatical finance, Monte Carlo method

1. Introduction

Let $(\Omega, \mathcal{F}, \{\mathcal{F}_t\}_{t\in[0,\infty)}, P)$ be a filtered space with the usual condition, and $\{B_t\}_{t\in[0,\infty)}$ be a d-dimensinal $\mathcal{F}_t$-Brownian motion. Let $T > 0$, and let $\sigma : [0,T]\times\mathbf{R}^D \to \mathbf{R}^D\otimes\mathbf{R}^d$ and $b : [0,T]\times\mathbf{R}^D \to \mathbf{R}^D$ be continuous functions. For each $s \in [0,T]$ and $x \in \mathbf{R}^D$, let $X(t;s,x)$, $t \in [s,T]$ be a solution of the following SDE.

$$X(t;s,x) = x + \int_s^t \sigma(r, X(r;s,x))dB_r + \int_s^t b(r, X(r;t,x))dr, \qquad t \in [s,T]. \tag{1}$$

We assume that the above SDE (1) has a path-wise unique solution for every $(s,x) \in [0,T]\times\mathbf{R}^D$.

Let $\tilde{\mathcal{S}}_s^t$, $0 \le s \le t \le T$, be the set of $\mathcal{F}_t$-stopping times τ with $s \le \tau \le t$. Let $g : [0,T]\times\mathbf{R}^D \to \mathbf{R}$ be a continuous function with suitable conditions. Then, concerning the pricing of American derivatives, we are interested in computing the following value function

$$u(s,x) = \sup\{E[g(\tau, X(\tau;s,x))]; \tau \in \tilde{\mathcal{S}}_s^T\}, (s,x) \in [0,T]\times\mathbf{R}^D.$$

There are several attempts to compute the value function u numerically. However, it seems that there are not so good method if D is not small. Let $N \geq 2$ and let T_n, $n = 0, 1, \ldots, N$, be positive numbers such that $0 = T_0 < T_1 < \ldots < T_N = T$. Let $\mathcal{S}_n$, $n = 0, 1, \ldots, N$, be the set of $\mathcal{F}_t$-stopping times taking value in $\{T_n, T_{n+1}, \ldots, T_N\}$. Concerning the pricing of Bermuda type derivatives, we are interested in computing the following value functions.

$$v_n(x) = \sup\{E[g(\tau, X(\tau; s, x))]; \tau \in \mathcal{S}_n\}, \qquad n = 0, 1, \ldots, N.$$

Let us define a probability measure $p_n(x, \cdot)$ over $\mathbf{R}^D$ for each $n = 0, 1, \ldots, N-1$, and $x \in \mathbf{R}^D$ by

$$p_n(x, A) = P(X(T_{n+1}; T_n, x) \in A), \qquad \text{for a Borel set } A \text{ in } \mathbf{R}^D,$$

and define an operator P_n, $n = 0, 1, \ldots, N-1$, by

$$P_n f(x) = \int_{\mathbf{R}^D} f(y) p_n(x, dy) = E[f(X(T_{n+1}; T_n, x))]$$

for a measurable function f on $\mathbf{R}^D$. Then v_n, $n = N, N-1, \ldots, 0$, are given inductively by the following.

$$v_N(x) = g(T_N, x),$$

$$v_{n-1}(x) = (P_{n-1} v_n)(x) \vee g(T_{n-1}, x).$$

So the value function $v_0(x)$ is easily given mathematically. However, if D is not small, it is not easy to memorize a function on $\mathbf{R}^D$, and so practically it is not easy to compute $v_0(x)$.

Several people suggest a Monte-Carlo method to compute the value function ([1], [2], [5], [6]). In this paper, we discuss the method given by Longstaff-Schwartz [4].

Let $W_n = \mathbf{R}^{(N+1-n)D}$, $n = 0, 1, \ldots, N$, and let $P_x^{(n)}$, $x \in \mathbf{R}^D$, be the distribution of $\{X(T_m; T_n, x)\}_{m=n}^N$ on W_n. Then $P_x^{(n)}$, $n = 0, 1, \ldots, N$, $x \in \mathbf{R}^D$, is a time inhomogeneous Markov chain on $\mathbf{R}^D$. Let ν_0 be a probability measure over $\mathbf{R}^D$. Let $L_n \geq 1$, $n = 0, 1, \ldots, N-1$, and $\mathbf{X}_{n,\ell} = \{X_{n,\ell}(m)\}_{m=0}^N$, $\ell = 1, \ldots, L_n$, $n = 0, 1, \ldots, N-1$, are identically independent random vectors defined on the probability measure $(\tilde{\Omega}, \tilde{\mathcal{F}}, \tilde{P})$ whose distribution is $P_{\nu_0}^{(0)} = \int_{\mathbf{R}^D} P_x^{(0)} \nu_0(dx)$. Let $K_n \geq 1$, $n = 0, 1, \ldots, N-1$, and $\psi_{n,k}$, $k = 1, \ldots, K_n$, $n = 0, 1, \ldots, N-1$, are functions on $\mathbf{R}^D$.

We assume the following assumption (A1).

(A1) D_n, $n = 0, 1, \ldots, N-1$, are measurable sets in $\mathbf{R}^N$ such that $(P_n v_{n+1})(x) \geq g(T_n, x)$ for any $x \in \mathbf{R}^D \setminus D_n$.

Remark 1 (1) $D_n = \mathbf{R}^D$ satisfies the assumption (A1).
(2) If $g(t,x) \geq 0$, for any $(t,x) \in [0,T] \times \mathbf{R}$, then $D_n = \{x \in \mathbf{R}^D; g(T_n, x) > 0\}$ satisfies the assumption (A1).

We define random functions $H_n : \mathbf{R}^D \times \tilde{\Omega} \to \mathbf{R}$, $n = N, N-1, \dots, 1, 0$, inductively by the following.

$$H_N(x) = 1.$$

When $\mathbf{H}_{n+1} = \{H_m\}_{m=n+1}^N$, are defined, we let

$$\sigma_{n,\ell} = \min\{m \geq n+1; H_m(X_{n,\ell}(m)) > 0\}, \qquad \ell = 1, \dots L_n,$$

and let $\{\tilde{a}_{n,k}\}_{k=1}^{K_n}$ be the minimizing point of the function

$$\begin{aligned} F_n(\{a_k\}_{k=1}^{K_n}) &= \frac{1}{L_n} \sum_{\ell=1}^{L_n} |g(T_{\sigma_{n,\ell}}, X_{n,\ell}(\sigma_{n,\ell})) \\ &\quad - \sum_{k=1}^{K_n} a_n \psi_{n,k}(X_{n,\ell}(n))|^2 1_{D_n}(X_{n,\ell}(n)). \end{aligned}$$

Finally we define H_n by

$$H_n(x) = \begin{cases} g(T_n, x) - \sum_{k=1}^{K_n} \tilde{a}_{n,k} \psi_{n,k}(x), & x \in D_n \\ -1, & x \in \mathbf{R}^D \setminus D_n. \end{cases}$$

Let $\tilde{v}_0 : \tilde{\Omega} \to \mathbf{R}$ be given by

$$\tilde{v}_0 = \frac{1}{L_0} \sum_{\ell=1}^{L_0} g(\sigma_{0,\ell}, X_{0,\ell}(\sigma_{0,\ell}))$$

In the present paper, we discuss the estimate on $E[(\tilde{v}_0 \vee g(0,x_0) - v_0(x_0))^2 \wedge 1]$ when $\nu_0(dx) = \delta_{x_0}(dx)$ (see Corollary 3.2).

2. Preliminary results

Let $W_n = \mathbf{R}^{(N+1-n)D}$, $n = 0, 1, \dots, N$, and let $P_x^{(n)}$, $x \in \mathbf{R}^D$, be the distribution of $\{X(T_m; T_n, x)\}_{m=n}^N$ on W_n. Then $P_x^{(n)}$, $n = 0, 1, \dots, N$, $x \in \mathbf{R}^D$, is a Markov chain on $\mathbf{R}^D$.

For any measurable function h on $\mathbf{R}^D$ and $n, m = 0, 1, \dots, N$ with $n \leq m$, let $\tau_m(\cdot; h) : W_n \to \{m, N\}$ by

$$\tau_m(w; h) = \begin{cases} m, & h(w(m)) > 0, \\ N, & h(w(m)) \leq 0. \end{cases}$$

Lemma 2 *Let* $h_n : \mathbf{R}^N \to \mathbf{R}$, $n = 0, 1, \ldots, N$, *be given, and assume that* $h_n(x) \le 0$, $x \in \mathbf{R}^N \setminus D_n$, *and that* $h_N(x) = 1$. *Let* $\sigma_n : W_n \to \{n, n+1, \ldots, N\}$ *be given by*

$$\sigma_n(w) = \sigma_n(w; \{h_m\}_{m=n}^{N-1}) = \bigwedge_{m=n}^{N-1} \tau_m(w; h_m), \qquad w \in W_n.$$

Moreover, let $u_n : \mathbf{R}^D \to \mathbf{R}$ *be given by*

$$u_n(x) = u_n(x; \{h_m\}_{m=n}^{N}) = E^{P_x^{(n)}}[g(T_{\sigma_n}, w(\sigma_n))], \qquad x \in \mathbf{R}^D,$$

Then we have the following.
(1) $|u_n(x) - v_n(x)| \le |P_n(u_{n+1} - v_{n+1})(x)| + 1_{D_n}(x)|P_n u_{n+1}(x) - (g(T_n, x) - h_n(x))|$ *for any* $n = 0, 1, \ldots, N-1$, *and* $x \in \mathbf{R}^D$.
(2) $|u_n(x) - v_n(x)| \le |P_n(u_{n+1} - v_{n+1})(x)| + 1_{D_n}(x) 1_{\{1\}}(sgn(P_n u_{n+1}(x) - g(T_n, x)) sgn(h_n(x))) |P_n u_{n+1}(x) - g(n, x)|$.

Here

$$sgn(a) = \begin{cases} 1, & a > 0, \\ 0, & a = 0, \\ -1, & a < 0. \end{cases}$$

Proof. Note that $u_n(x) \le v_n(x)$, for all $n = 0, 1, \ldots, N-1$, and $x \in \mathbf{R}^D$. Let $\tilde{u}_n(x) = g(T_n, x) - h_n(x)$, $x \in \mathbf{R}^D$.

Let $n = 0, 1, \ldots, N-1$, and $x \in \mathbf{R}^D$, and fix them for a while.
Case 1. Suppose that $h_n(x) > 0$.
Then we see that $x \in D_n$ and $g(T_n, x) > \tilde{u}_n(x)$. So we have

$$v_n(x) = g(T_n, x) + (P_n v_{n+1}(x) - g(T_n, x)) \vee 0 \le g(T_n, x) + |P_n v_{n+1}(x) - \tilde{u}_n(x)|.$$

This implies

$$g(T_n, x) \ge v_n(x) - |P_n(v_{n+1} - u_{n+1})(x)| - |P_n u_{n+1}(x) - \tilde{u}_n(x)|.$$

Case 2. Suppose that $h_n(x) \le 0$, and $x \in D_n$.
Then we see that $g(T_n, x) \le \tilde{u}_n(x)$. So we see that

$$\begin{aligned} v_n(x) &\le P_n v_{n+1}(x) \vee \tilde{u}_n(x) \\ &\le P_n u_{n+1}(x) + |P_n(v_{n+1} - u_{n+1})(x)| + |P_n u_{n+1}(x) - \tilde{u}_n(x)|. \end{aligned}$$

Case 3. Suppose that $h_n(x) \le 0$, and $x \in \mathbf{R}^D \setminus D_n$.
Then we see that $g(T_n, x) \le (P_n v_{n+1})(x)$. So we have

$$v_n(x) = P_n v_{n+1}(x) \le P_n u_{n+1}(x) + |P_n(u_{n+1} - v_{n+1})(x)|.$$

So we see that for any $n = 0, 1, \ldots, N-1$,

$$u_n = 1_{\{h_n>0\}} g(T_n, \cdot) + 1_{\{h_n \le 0\}}(P_n u_{n+1})$$

$$\ge 1_{\{h_n>0\}}(v_n - |P_n(v_{n+1} - u_{n+1})| - |P_n u_{n+1} - \tilde{u}_n|)$$

$$+1_{\{h_n \le 0\}} 1_{D_n}(v_n - |P_n(v_{n+1} - u_{n+1})| - |P_n u_{n+1} - \tilde{u}_n|)$$

$$+1_{\{h_n \le 0\}} 1_{\mathbf{R}^D \setminus D_n}(v_n - |P_n(v_{n+1} - u_{n+1})|).$$

Thus we see that

$$0 \le v_n - u_n \le |P_n(v_{n+1} - u_{n+1})| + |1_{D_n} P_n u_{n+1} - \tilde{u}_n|.$$

This implies the assertion (1).

Now let us prove the assertion (2). Let ξ is a positive measurable function on $\mathbf{R}^D$. Since $\tau_n(w; \xi h_n) = \tau_n(w; h_n)$, we see from the assertion (1) that

$$\begin{aligned} |u_n(x) - v_n(x)| \ \le \ & |P_n(u_{n+1} - v_{n+1})(x)| \\ & +1_{D_n}(x)|P_n u_{n+1}(x) - g(T_n, x) + \xi(x) h_n(x)|. \end{aligned}$$

Noting that

$$\inf\{a + tb;\ t > 0\} = 1_{\{1\}}(sgn(a)sgn(b))|a|, \qquad a, b \in \mathbf{R},$$

we have the assertion (2).

This completes the proof. □

Let ν_0 be a probability measure on $\mathbf{R}^D$ and define probability measures ν_n, $n = 1, \ldots, N$, inductively by

$$\nu_{n+1}(dx) = \int_{\mathbf{R}^D} p_n(y; dx)\nu_n(dy), \qquad n = 0, 1, \ldots, N-1.$$

Then we have the following as an easy consequence of Lemma 2.1.

Corollary 3 *Let h_n and u_n be the same as the previous lemma. Then we have the following.*

$$\begin{aligned} (\int_{\mathbf{R}^D} |u_n(x) - v_n(x)|^2 \nu_n(dx))^{1/2} \le |(\int_{\mathbf{R}^D} |u_{n+1} - v_{n+1}(x)|^2 \nu_{n+1}(dx))^{1/2} \\ + \int_{D_n} |P_n u_{n+1}(x) - (g(T_n, x) - h_n(x))|^2)^{1/2} \end{aligned}$$

for any $n = 0, 1, \ldots, N-1$.

3. Main result

Let us think of the situation in Introduction.

We assume the following additionally to (A-1).

(A2) $\psi_{n,k}$, $k = 1,\dots,K_n$, is linear ly independent in $L^2(D_n; d\nu_n)$, $n = 0,1,\dots,N-1$, where ν_n is the probability law of $w(n)$ under $P^{(0)}_{\nu_0}(dw)$.

(A3) $\int_{D_n} \psi_{n,k}(x)^4 \nu_n(dx) < \infty\ k = 1,\dots,K_n, n = 0,1,\dots,N-1,$

and

$$\int_{\mathbf{R}^D} E^{P^{(0)}_x}[(\sum_{m=1}^{N} g(T_n, w(T_n))^4]\nu_0(dx) < \infty, \qquad n = 0,1,\dots,N.$$

Let $U_n(x) = u_n(\cdot;\{H_m\}_{m=n}^N))(x)$. Here u_n is as in Lemma 2.1. Let $\bar{a}_n = \{\bar{a}_{n,k}\}_{k=1}^{K_n}$ be the minimizing point of the function

$$\bar{F}_n(\{a_k\}_{k=1}^{K_n}) = \int_{D_n} |(P_n U_{n+1})(x) - \sum_{k=1}^{K_n} a_n \psi_{n,k}(x))|^2 \nu_n(dx).$$

Then we have the following.

Theorem 4 (1) *There is a constant $C > 0$ such that*

$$E^{\tilde{P}}[1 \wedge (\sum_{k=1}^{K_n} |\tilde{a}_{n,k} - \bar{a}_{n,k}|^2)] \leq \frac{C}{L_n}.$$

(2)

$$\begin{aligned}
&(\int_{\mathbf{R}^D} |U_n(x) - v_n(x)|^2 \nu_n(dx))^{1/2} \\
&\quad \leq (\int_{\mathbf{R}^D} |U_{n+1}(x) - v_{n+1}(x)|^2 \nu_{n+1}(dx))^{1/2} \\
&\quad\quad + (\int_{\mathbf{R}^D} (\sum_{k=1}^{K_n} (\tilde{a}_{n,k} - \bar{a}_{n,k})\psi_{n,k}(x))^2 \nu_n(dx))^{1/2} \\
&\quad\quad + \inf\{(\int_{D_n} |(P_n U_{n+1})(x) - \sum_{k=1}^{K_n} a_k \psi_{n,k}(x)|^2 \nu_n(dx))^{1/2};\ a_k \in \mathbf{R}, \\
&\quad k = 1,\dots,K_n\}.
\end{aligned}$$

Proof. Let $\mathcal{I}_n$, $n = 0,1,\dots,N-1$, be the σ-algebra generated by $\mathbf{X}_{n,\ell}$, $\ell = 1,\dots,L_n$, and let $\mathcal{B}_n$, $n = 0,1,\dots,N-1$, be the σ-algebra generated by

$\cup_{m=n}^{N-1}\mathcal{I}_m$. Inductively, we see that H_n is $\mathcal{B}_n$-measurable, $n = N-1, N-2, \ldots, 0$. Also, we have

$$F_n(\{a_k\}_{k=1}^{K_n}) = \sum_{k,k'=1}^{K_n} C_{n,k,k'}^{(2)} a_k a_{k'} - 2\sum_{k=1}^{K_n} c_{n,k}^{(1)} a_k + C_n^{(0)},$$

where

$$C_{n,k,k'}^{(2)} = \frac{1}{L_n}\sum_{\ell=1}^{L_n}(1_{D_n}\psi_{n,k}\psi_{n,k'})(X_{n,\ell})$$

$$c_{n,k}^{(1)} = \frac{1}{L_n}\sum_{\ell=1}^{L_n}(1_{D_n}\psi_{n,k})(X_{n,\ell})g(T_{\sigma_{n,\ell}}, X_{n,\ell}(\sigma_{n,\ell})).$$

Note that $\sigma_{n,\ell} = \sigma_{n+1}(\mathbf{X}_{n,\ell}(\cdot); \{H_m\}_{m=n+1}^N)$. Therefore we have

$$\bar{C}_{n,k,k'}^{(2)} = E^{\tilde{P}}[C_{n,k,k'}^{(2)}|\mathcal{B}_{n+1}] = \int_{D_n} \psi_{n,k}(x)\psi_{n,k'}(x)\nu_n(dx), \tag{2}$$

and

$$\bar{c}_{n,k}^{(1)} = E^{\tilde{P}}[c_{n,k}^{(1)}|\mathcal{B}_{n+1}] = \int_{D_n} \psi_{n,k}(x)(P_nU_{n+1})(x)\nu_n(dx). \tag{3}$$

Let $R_{n,k,k'}^{(2)} = C_{n,k,k'}^{(2)} - \bar{C}_{n,k,k'}^{(2)}$, and $r_{n,k}^{(1)} = c_{n,k}^{(1)} - \bar{r}_{n,k}^{(1)}$. Also, let $C_n^{(2)} = \{C_{n,k,k'}^{(2)}\}_{k,k'=1}^D$, $\bar{C}_n^{(2)} = \{\bar{C}_{n,k,k'}^{(2)}\}_{k,k'=1}^D$, $R_n^{(2)} = \{R_{n,k,k'}^{(2)}\}_{k,k'=1}^D$, $c_n^{(1)} = \{c_{n,k}^{(1)}\}_{k=1}^D$, $\bar{c}_n^{(1)} = \{\bar{c}_{n,k}^{(1)}\}_{k=1}^D$, and $r_n^{(1)} = \{r_{n,k}^{(1)}\}_{k=1}^D$. Then $C_n^{(2)}$ and$R_n^{(2)}$ are are $D \times D$ random matrices, $c_n^{(1)}$, $\bar{c}_n^{(1)}$ and $r_n^{(1)}$ are D-dimesional random vectors, and $\bar{C}_n^{(2)}$ is a nonrandom $D \times D$ matrix.

Then we see that

$$\tilde{a}_n = C_n^{(2)-1}c_n^{(1)}, \qquad \bar{a}_n = \bar{C}_n^{(2)-1}\bar{c}_n^{(1)}, \qquad n = 0, \ldots, N-1.$$

Also, we see that

$$E^{\tilde{P}}[(R_{n,k,k'}^{(2)})^2]$$

$$= \frac{1}{L_n}E[Var[1_{D_n}(X_{n,1}(n))\psi_{n,k}(X_{n,1}(n))\psi_{n,k'}(X_{n,1}(n))|\mathcal{B}_{n+1}]]$$

$$\leq \frac{1}{L_n}\int_{D_n} \psi_{n,k}(x)^2\psi_{n,k'}(x)^2\nu_n(dx),$$

and

$$E^{\tilde{P}}[(r_{n,k}^{(1)})^2]$$

$$= \frac{1}{L_n}E[Var[1_{D_n}(X_{n,1}(n))\psi_{n,k}(X_{n,1}(n))g(\sigma_{n,1}, X_{n,1}(\sigma_{n,1}))|\mathcal{B}_{n+1}]]$$

$$\leq \frac{1}{L_n}\int_{D_n}\psi_{n,k}(x)^2 E^{P_x^{(n)}}[g(T_{\sigma_{n+1}(w;\{H_m\}_{m=n+1}^N)}) \\ w(\sigma_{n+1}(w;\{H_m\}_{m=n+1}^N)))^2]\nu_n(dx).$$

So we have

$$|\bar{c}_n^{(1)}| \leq b_n^{(1)},$$

where

$$\begin{aligned} b_n^{(1)} &= (\int_{D_n}(\sum_{k=1}^{K_n}\psi_{n,k}(x)^2)\nu_n(dx))^{1/2} \\ &\quad (\int_{\mathbf{R}^D} E^{P_x^{(0)}}[(\max_{m=1,\dots,N} g(T_m, w(T_m))^2]\nu_0(dx))^{1/2}. \end{aligned}$$

also, we have

$$E^{\tilde{P}}[\| R_n^{(2)} \|^2] \leq \frac{1}{L_n}\int_{D_n}(\sum_{k=1}^{K_n}\psi_{n,k}(x)^2)^2\nu_n(dx) \tag{4}$$

and

$$\begin{aligned} E^{\tilde{P}}[|r_n^{(1)}|^2] &\leq \frac{1}{L_n}(\int_{D_n}(\sum_{k=1}^{K_n}\psi_{n,k}(x)^2)^2\nu_n(dx))^{1/2} \\ &\quad (\int_{\mathbf{R}^D} E^{P_x^{(0)}}[(\max_{m=1,\dots,N} g(T_m, w(T_m))^4]\nu_0(dx))^{1/2}. \end{aligned} \tag{5}$$

If $\| \bar{C}_n^{(2)-1} R_n^{(2)} \| \leq 1/2$, we have

$$\| (\bar{C}_n^{(2)} + R_n^{(2)})^{-1} - \bar{C}_n^{(2)-1} \| \\ = \| ((I + \bar{C}_n^{(2)-1} R_n^{(2)})^{-1} - I)\bar{C}_n^{(2)-1} \| \leq 2 \| \bar{C}_n^{(2)-1} \| \| R_n^{(2)} \| .$$

Here $\| \cdot \|$ is the operator norm of a matrix. So if $\| \bar{C}_n^{(2)-1} \| \| R_n^{(2)} \| \leq 1/2$ and $|r_n^{(1)}| \leq 1$, we have

$$|\tilde{a}_n - \bar{a}_n| = |((\bar{C}_n^{(2)} + R_n^{(2)})^{-1} - \bar{C}_n^{(2)-1})(\bar{c}_n^{(1)} + r_n^{(1)}) + \bar{C}_n^{(2)-1} r_n^{(1)}| \\ \leq 2 \| \bar{C}_n^{(2)-1} \| \| R_n^{(2)} \| (|\bar{c}_n^{(1)}| + 1) + \| \bar{C}_n^{(2)-1} \| |r_n^{(1)}|.$$

So we have

$$E^{\tilde{P}}[|\tilde{a}_n - \bar{a}_n|^2 \wedge 1] \\ \leq E^{\tilde{P}}[|\tilde{a}_n - \bar{a}_n|^2, \| \bar{C}_n^{(2)-1} \| \| R_n^{(2)} \| \leq 1/2, |r_n^{(1)}| \leq 1] \\ + \tilde{P}(\| \bar{C}_n^{(2)-1} \| \| R_n^{(2)} \| > 1/2) + \tilde{P}(|r_n^{(1)}| > 1)$$

$$\leq (8(b_n^{(1)}+1)^2 \parallel \bar{C}_n^{(2)-1} \parallel^4 +4 \parallel \bar{C}_n^{(2)-1} \parallel^{-2})$$

$$E^{\tilde{P}}[\parallel R_n^{(2)} \parallel^2] + (2 \parallel \bar{C}_n^{(2)-1} \parallel^2 +1)E^{\tilde{P}}[|r_n^{(1)}|^2].$$

This and Equations (4) and (5) imply the assertion (1).
The assertion (2) is an easy consequence of Lemma 2.1. □

Let $V_n = \sum_{k=1}^{K_n} \mathbf{R}\psi_{n,k} \subset L^2(\mathbf{R}^D; d\nu_n)$, $n = 0, 1, \ldots, N-1$. Then it is easy to see that U_n's are determined by $\mathbf{X}_{n,\ell}$, $\ell = 1, \ldots, L_n$, $n = 0, \ldots, N$ and V_n's , and are independent of a choice of bases $\{\psi_{n,k}\}_{k=1}^{K_n}$. Let

$$d_n = \inf\{(\int_{\mathbf{R}^D} (\sum_{k=1}^{K_n} \psi_k(x)^2)^2 \nu_n(dx))^{1/2};\ \{\psi_k\}_{k=1}^{K_n} \text{ is a orthogonal basis of } V_n\},$$

and

$$c_0 = (\int_{\mathbf{R}^D} E^{P_x^{(0)}}[(\max_{m=1,\ldots,N} g(T_m, w(T_m))^4]\nu_0(dx))^{1/4}.$$

Then we have the following from the proof of Theorem 3.1.

Corollary 5 $E[(\int_{\mathbf{R}^D} |U_n(x) - v_n(x)|^2 \nu_n(dx)) \wedge 1]^{1/2}$

$$\leq E[(\int_{\mathbf{R}^D} |U_{n+1}(x) - v_{n+1}(x)|^2 \nu_{n+1}(dx)) \wedge 1]^{1/2} + 8L_n^{-1/2} d_n (K_n^{1/2} c_0^{1/2} + 1)$$

$$+E[\inf\{(\int_{D_n} |(P_n U_{n+1})(x) - \psi(x)|^2 \nu_n(dx));\ \psi \in V_n\} \wedge 1]^{1/2}.$$

Then we have the following.

Corollary 6 *Let $\tilde{v}_0$ be as in Introduction, and assume that $\nu_0 = \delta_{x_0}$ for some $x_0 \in \mathbf{R}^D$. Then we have*

$$E[(\tilde{v}_0 \vee g(0, x_0) - v_0(x_0))^2 \wedge 1]^{1/2} \leq E_1 + E_2,$$

where

$$E_1 = L_0^{-1/2} c_0 + \sum_{n=1}^{N-1} 8(L_n)^{-1/2} d_n (K_n^{1/2} c_0^{1/2} + 1)$$

and

$$E_2 = \sum_{n=1}^{N-1} E[\inf\{(\int_{D_n} |(P_n U_{n+1})(x) - \psi(x)|^2 \nu_n(dx));\ \psi \in V_n\} \wedge 1]^{1/2}.$$

Proof. Note that $v_0(x_0) = (P_1 v_1)(x_0) \vee g(0, x_0)$. We also see that

$$E[g(T_{\sigma_{0,\ell}}, X_{0,\ell}(\sigma_{0,\ell}))|\mathcal{B}_1] = (P_1 U_1)(x_0).$$

So we see that

$$E[(\tilde{v}_0 - (P_1 U_1)(x_0))^2] \leq L_0^{-1/2} c_0.$$

Since

$$|(\tilde{v}_0 \vee g(0, x_0) - v_0(x_0)| \leq |\tilde{v}_0 - (P_1 U_1)(x_0)| + |(P_1 U_1)(x_0) - (P_1 v_1)(x_0)|$$

$$\leq |\tilde{v}_0 - (P_1 U_1)(x_0)| + (\int_{\mathbf{R}^D} |U_1(x) - v_1(x)|^2 \nu_1(dx))^{1/2},$$

we have our assertion from Corollary 3.1.

This completes the proof. □

Remark 7 We may call E_1 and E_2 in Corollary 3.2 the simulation error and the approximation error, since E_1 is an error caused by Monte Carlo simulation and E_2 is an error caused by the approximation by systems of functions. E_1 tends to zero as L_n tends to infinity.

We will study how small E_2 is in the special case in the next section.

4. Relation to hypoellipticity

Let $V_0, V_1, \ldots, V_d \in C_b^\infty(\mathbf{R}^D; \mathbf{R}^D)$. Here $C_b^\infty(\mathbf{R}^N; \mathbf{R}^n)$ denotes the space of $\mathbf{R}^n$-valued smooth functions defined in $\mathbf{R}^N$ whose derivatives of any order are bounded. We regard elements in $C_b^\infty(\mathbf{R}^D; \mathbf{R}^D)$ as vector fields on $\mathbf{R}^D$.

Now let $X(t, x)$, $t \in [0, \infty)$, $x \in \mathbf{R}^D$, be the solution to the Stratonovich stochastic integral equation

$$X(t, x) = x + \sum_{i=0}^{d} \int_0^t V_i(X(s, x)) \circ dB^i(s). \tag{6}$$

Then there is a unique solution to this equation. We think of the SDE (6) instead of the SDE (1) as the SDE for the underlying stochastic process.

Let $\mathcal{A} = \{\emptyset\} \cup \bigcup_{k=1}^\infty \{0, 1, \ldots, d\}^k$ and for $\alpha \in \mathcal{A}$, let $|\alpha| = 0$ if $\alpha = \emptyset$, let $|\alpha| = k$ if $\alpha = (\alpha^1, \ldots, \alpha^k) \in \{0, 1, \ldots, d\}^k$, and let $\| \alpha \| = |\alpha| + \mathrm{card}\{1 \leq i \leq |\alpha|;\ \alpha^i = 0\}$. Let $\mathcal{A}_0$ and $\mathcal{A}_1$ denote $\mathcal{A} \setminus \{\emptyset\}$ and $\mathcal{A} \setminus \{\emptyset, 0\}$, respectively. Also, for each $m \geq 1$, $\mathcal{A}(m)$, $\mathcal{A}_0(m)$ and $\mathcal{A}_1(m)$ denote $\{\alpha \in \mathcal{A};\ \| \alpha \| \leq m\}$, $\{\alpha \in \mathcal{A}_0;\ \| \alpha \| \leq m\}$ and $\{\alpha \in \mathcal{A}_1;\ \| \alpha \| \leq m\}$ respectively.

We define vector fields $V_{[\alpha]}$, $\alpha \in \mathcal{A}$, inductively by

$$V_{[\emptyset]} = 0, \qquad V_{[i]} = V_i, \quad i = 0, 1, \ldots, d,$$

$$V_{[\alpha i]} = [V_{[\alpha]}, V_i], \qquad i = 0, 1, \ldots, d.$$

Definition 8 *We say that a system* $\{V_i; i = 0, 1, \ldots, d\}$ *of vector fields satisfies the condition* (UFG), *if there are an integer* ℓ *and* $\varphi_{\alpha,\beta} \in C_b^\infty(\mathbf{R}^D)$, $\alpha \in \mathcal{A}_1$, $\beta \in \mathcal{A}_1(\ell)$, *satisfying the following.*

$$V_{[\alpha]} = \sum_{\beta \in \mathcal{A}_1(\ell)} \varphi_{\alpha,\beta} V_{[\beta]}, \qquad \alpha \in \mathcal{A}_1.$$

We assume that a system $\{V_i; i = 0, 1, \ldots, d\}$ of vector fields satisfies the condition (UFG) throughout, and let ℓ be as in Definition 4.1.. Let $\lambda_0 : \mathbf{R}^D \to [0, \infty)$ be a continuous function given by

$$\lambda_0(x) = \inf\{ \sum_{\alpha \in \mathcal{A}_1(\ell)} (V_{[\alpha]}(x), \xi)^2;\ \xi \in \mathbf{R}^D,\ |\xi| = 1\} \qquad x \in \mathbf{R}^D.$$

Let us define a semigroup of linear operators $\{P_t\}_{t \in [0,\infty)}$ by

$$(P_t f)(x) = E[f(X(t,x))], \qquad t \in [0, \infty),\ f \in C_b(\mathbf{R}^D).$$

Now let ν_0 be a probability measure on $\mathbf{R}^D$ and let ν_s, $s \in (0, \infty)$, be a probability measure on $\mathbf{R}^D$ given by

$$\nu_t(A) = \int_{\mathbf{R}^D} \nu_0(dx) P(X(t,x) \in A),$$

for any Borel set A in $\mathbf{R}^D$.

We assume the following, moreover.

(AM) (1) There is an $\varepsilon > 0$ such that

$$\int_{\mathbf{R}^D} \exp(|x|^{(1+\varepsilon)}) \nu_0(dx) < \infty.$$

(2) For any $p \in (1, \infty)$

$$\int_{\mathbf{R}^D} \lambda_0(x)^{-p} \nu_0(dx) < \infty.$$

Let $\lambda : \mathbf{R}^D \to [0, \infty)$ be given by

$$\lambda(x) = \begin{cases} (trace A(x)^{-1})^{-1}, & \text{if } \lambda_0(x) > 0, \\ 0, & \text{if } \lambda_0(x) = 0. \end{cases}$$

Then we can easily see that

$$N^{-1} \lambda_0(x) \le \lambda(x) \le \lambda_0(x), \qquad x \in \mathbf{R}^N.$$

Also, we have the following (see [3]).

Proposition 9 (1) *For any* $n \geq 1$ *and* $i_1, \ldots, i_n \in \{1, \ldots, N\}$, *there is a* $C > 0$ *such that*

$$|\frac{\partial^n}{\partial x^{i_1} \cdots \partial x^{i_n}} \lambda(x)| \leq C\lambda(x)^{-(n-1)}, \qquad x \in \mathbf{R}^N.$$

(2) *For any* $n \geq 1$ *and* $i_1, \ldots, i_n \in \{1, \ldots, D\}$ *there is a constant* $C > 0$ *such that*

$$\| \lambda_0^{2n-1} \frac{\partial^n}{\partial x^{i_1} \ldots \partial x^{i_n}} P_t f \|_\infty \leq Ct^{-n\ell/2} \| f \|_\infty$$

for any bounded measurable function $f : \mathbf{R}^N \to \mathbf{R}$ *and* $t \in (0, 1]$.
(3) *For any* $p \in (1, \infty)$ *and* $s > 0$,

$$\int_{\mathbf{R}^D} \lambda_0(x)^{-p} \nu_s(dx) < \infty.$$

Let $\mathcal{P}_n$, $n \geq 1$, be the set of polynomials on $\mathbf{R}^D$ of degree at most n. Then we have the following.

Theorem 10 (1) *Let* $\delta = (\varepsilon \wedge 1)/2$. *Then*

$$\int_{\mathbf{R}^D} \exp(|x|^{(1+\delta)}) \nu_s(dx) < \infty, \qquad s \geq 0.$$

(2) *For each* $t, s \geq 0$, $n \geq 1$, *and a bounded mesurable function* $f : \mathbf{R}^D \to \mathbf{R}$ *let*

$$d_n(f; t, s) = \inf\{(\int_{\mathbf{R}^D} |P_t f(x) - g(x)|^2 \nu_s)^{1/2};\ g \in \mathcal{P}_n\}.$$

Also, let

$$d_n(t, s) = \sup\{d_n(f; t, s);\ f \text{ is a bounded mesurable function with } |f| \leq 1\}.$$

Then $n^p d_n(t, s) \to 0$, *as* $n \to \infty$, *for any* $t > 0$ *and* $s \geq 0$.

Proof. We need some preparations to prove Theorem 4.1. The assertion (1) is obvious, since there is a constant $\alpha > 0$ for each $s > 0$ such that

$$\sup\{E[\exp(\alpha |X(s, x) - x|^2)];\ x \in \mathbf{R}^D\} < \infty.$$

Let us prove the assertion (2). First we fix $t > 0$ and $s \geq 0$. Let $\mathcal{S}$ be the set of $f \in C_b^\infty(\mathbf{R}^D)$ with $\| f \|_\infty \leq 1$.

Let $\psi \in C^\infty(\mathbf{R})$ be such that $0 \leq \psi \leq 1$, $\psi(z) = 1$, $z \leq 1$, and $\psi(z) = 0$, $z \geq 2$. Also, let $\varphi_r \in C_0^\infty(\mathbf{R}^D)$, $r \geq 1$, be given by

$$\varphi_r(x) = \psi(r^{-1}|x|)(1 - \psi(r\lambda(x)), \qquad x \in \mathbf{R}^D.$$

Then for each $k \geq 0$ there is a constant $C_{1,k}$ such that

$$|\nabla^k \varphi_r(x)| \leq C_{1,k} r^k 1_{(1,\infty)}(r\lambda(x)) 1_{[0,2r)}(|x|), \qquad x \in \mathbf{R}^D.$$

So we see that for each $k \geq 0$ there is a constant $C_{2,k}$ such that

$$|\nabla^k (\varphi_r P_t f)(x)| \leq C_{2,k} r^k 1_{[0,2r)}(|x|), \qquad x \in \mathbf{R}^D \tag{7}$$

for any $f \in \mathcal{S}$.

Also from the assetion (1), we see that

$$\int_{\mathbf{R}^D} |1 - \varphi_r(x)| \nu_s(dx) = o(r^{-p}), \qquad r \to \infty \tag{8}$$

for any $p \in [1, \infty)$.

Let

$$\rho_r(\xi; f) = (2\pi)^{-D} \int_{\mathbf{R}^D} \exp(-ix \cdot \xi) \varphi_r(x) (P_t f)(x) dx,$$

$$\xi \in \mathbf{R}^D, \ r \geq 1, \ f \in C_b^\infty(\mathbf{R}^D).$$

Thus we have

$$\varphi_r(x)(P_t f)(x) = \int_{\mathbf{R}^D} \exp(ix \cdot \xi) \rho_r(\xi; f) d\xi.$$

Then by Equation (7), we see that for each $k \geq 0$ there is a constant $C_{3,k}$ such that

$$|\rho_r(\xi; f)| \leq C_{3,k} r^{D+2k} (1 + |\xi|^2)^{-k} \tag{9}$$

for any $f \in \mathcal{S}$. Let $\gamma = \delta/4$,

$$g_{r,n}(x; f) = \int_{\mathbf{R}^D} \exp(ix \cdot \xi) 1_{[0,n^{2\gamma}]}(|\xi|)) \rho_r(\xi; f) d\xi$$

and

$$p_{r,n}(x; f) = \sum_{k=0}^{n-1} \int_{\mathbf{R}^D} \frac{(ix \cdot \xi)^k}{k!} 1_{[0,n^{2\gamma}]}(|\xi|)) \rho_r(\xi; f) d\xi$$

Then $p_{r,n}(x; f)$ is a polynomial of order $n - 1$. So we see that for each $k \geq 0$ there is a constant $C_{4,k}$ such that

$$\sup\{|\varphi_r(x)(P_t f)(x) - g_{r,n}(x; f)|; \ x \in \mathbf{R}^D\} \leq C_{4,k} r^{3D+2k} n^{-4k\gamma} \tag{10}$$

for any $f \in \mathcal{S}$.

Note that

$$|\exp(iz) - \sum_{k=0}^{n-1} \frac{(iz)^k}{k!}| \leq \frac{|z|^n}{n!}, \qquad z \in \mathbf{R}.$$

So we see that

$$|g_{r,n}(x;f) - p_{r,n}(x;f)| \leq \int_{\mathbf{R}^D} \frac{|x\cdot\xi|^n}{n!} 1_{[0,n^{2\gamma}]}(|\xi|)|\rho_r(\xi;f)|d\xi$$
$$\leq C_{3,0} r^D \frac{n^{D+2\gamma n}}{n!}|x|^n$$

for any $f \in \mathcal{S}$. Note that

$$\begin{aligned}(\int_{\mathbf{R}^D} |x|^{2n}\nu_s(dx))^{1/2} &\leq (\int_{\mathbf{R}^D} |x|^{2(1+\delta)n}\nu_s(dx))^{1/(2(1+\delta))} \\ &\leq C_4((2n)!)^{1/(2(1+\delta))},\end{aligned}$$

where $C_4 = (\int_{\mathbf{R}^D} \exp(|x|^{1+\delta})\nu_s(dx))^{1/(2(1+\delta))}$. Then we see that

$$(\int_{\mathbf{R}^D} |g_{r,n}(x;f) - p_{r,n}(x;f)|^2 \nu_s(dx))^{1/2}$$
$$\leq C_4 C_{3,0}(rn)^D \frac{n^{2\gamma n}}{n!}((2n)!)^{1/(2(1+\delta))}.$$

Since $2\gamma + (1+\delta)^{-1} < 1$, we see that

$$\sup\{(\int_{\mathbf{R}^D} |g_{r,n}(x;f) - p_{r,n}(x;f)|^2\nu_s(dx))^{1/2};\ f \in \mathcal{S}\} = o(n^{-p}),$$
$$n \to \infty \qquad (11)$$

for any $p \in (1,\infty)$.

Therefore Equations (7), (9), (10) and (11) imply our assertion.

References

1. Barraquand, J., Martineau, D.: Numerical valuation of high dimensional multivariate American securities. J. Financial and Quantitative Analysis **30**, 383-405 (1995)
2. Broadie, M., Glasserman, P.: Pricing American-style securities using simulation, J. Economical Dynamics and Control **21**, 1323-1352 (1997)
3. Kusuoka, S.: Malliavin Calculus revisited (in preparation)
4. Longstaff, F., Schwartz, E.: Valueing American Options by Simulation: A simple Least-Squares Approach. The Review of Financial Studies **14**, 113-147 (2001)
5. Tilley, J.A.: Valuing American options in a path simulation model. Transactions of the Society of Actuaries **XLV**, 499-520 (1993)
6. Tsitsiklis, J., Van Roy, B.: Optimal stoppimg of Markov processes: Hilbert space theory, approximation algorithms, and an application to pricing high dimensional financial derivatives, IEEE Transactions on Automatic Cotrol **10**, 1840-1851 (1999)

Adv. Math. Econ. 5, 167–195 (2003)

Advances in
MATHEMATICAL
ECONOMICS

Mathematical economics in Vienna between the Wars *

Isao Mutoh

National Defense Academy, School of Humanities and Social Sciences, Department of Public Policy, 1-10-20, Hashirimizu, Yokosuka-shi, Kanagawa 239-8686, Japan
(e-mail: mutoh@nda.ac.jp)

Received: November 9, 2001
Revised: July 22, 2002

JEL classification: B00, B21

Mathematics Subject Classification (2000): 01A60

Abstract. Wald was the first who solved the existence problem in economics. His paper should be regarded as a starting point of the axiomatization in economics. The movement toward an axiomatization in economics took place only in Vienna. We must not dismiss the intellectual backgrounds in Vienna, which could be classified under the following headings; (a) the research activities in economics, (b) Hilbert's philosophy on the foundations of mathematics, (c) the Vienna Circle's philosophy of science, and (d) the development of convex analysis. In this paper, we shall examine how the interplay of these factors stimulated Wald to the existence proof.

Key words: Abraham Wald, existence proof, the Vienna Circle, Hilbert's formalism, convex analysis

1. Introduction

Abraham Wald was the first who solved the existence problem in economics [53]. After a couple of years, John von Neumann succeeded in the existence proof of the balanced growth of a multi-sectoral economy [50]. The publication of these two papers should be regarded as a starting point of the axiomatization in economics. The papers by Wald and von Neumann were both published in the *Ergebnisse eines Mathematischen Kolloquiums*, which was an annual report of the colloquium organized by Karl Menger. He was a mathematician in the

* I am grateful to the anonymous referees for their helpful comments.

field of geometry and a professor at the University of Vienna. Wald intended to enter the University of Vienna to study geometry under the direction of Menger. Menger soon appreciated Wald's ability and invited him to his colloquium.

Among other outstanding mathematicians of the days who participated in this colloquium, we must mention such big names as Kurt Gödel, Hans Hahn, Georg Nöbeling, Franz Alt and Stafford Beer. Wald, Alt, Nöbeling and Gödel later became co-editors of the *Ergebnisse*. In addition, Alfread Tarski, Bronislaw Knaster, Karol Borsuk and Stefan Mazurkiewicz were frequent guests. And von Neumann dropped in the colloquium when he passed through Vienna during his summer journey to Budapest. In fact, Menger's colloquium constituted a sparkling constellation of big stars in the universe of mathematics.

The movement toward an axiomatization in economics took place only in Vienna. This was not the result of pure chance. We must not dismiss the intellectual backgrounds in Vienna, which produced and promoted this movement. They might be classified under the following headings; (a) the research activities in economics, (b) Hilbert's philosophy on the foundations of mathematics, (c) the new trend in the philosophy of science, and (d) the development of convex analysis. Without a combination of these factors, the movement toward an axiomatization in economics would not have emerged. It was the interplay of these factors that stimulated Wald to the existence proof. At this point, Karl Menger played an active role as an organizer. He built the bridge between these factors. He was an exceptional person who was well acquainted with mathematics, philosophy of science and economics[1].

Menger became deeply interested in social sciences, in economics in particular. In 1923, he edited the second edition of his father's *Grundsätze der Volkswirtschaftslehre* and wrote an introduction to it.

In his papers on the law of returns, Menger [28] examined alleged proofs of the law available in the economic literature. His papers were characterized by himself as a study in *meta-economics*[2]. He also participated in the Vienna Circle and was fully aware of the new trend in the philosophy of science. Without him, mathematical economics would not have flourished in Vienna.

In what follows, we are going to examine each of these factors and to clarify how the interplay of these factors motivated Wald to achieve the existence proof. In section 2, we give a brief sketch of the research activities in economics between the Wars. We explore the reason why the members in Menger's colloquium accepted the Walras-Cassel general equilibrium theory as a material of the discussion in the seminar.

---

[1] For more on K. Menger's contribution to social sciences, see Golland, L. and K. Sigmund [16].

[2] Menger ([28], p.280) said; "Following a suggestion of Hilbert, modern logicians refer to the study of the logical relations between the statements of a theory as the corresponding *meta-theory*. In this terminology, the contents of the present paper can be described as a chapter in *meta-economics*".

Wald axiomatized the Walras-Cassel system and gave an existence proof to it. His primary concern for the existence problem reflected Hilbert's philosophy on the foundations of mathematics. Hilbert insisted that the consistency of an axiomatic system was equivalent to the existence of the corresponding mathematical concept.

Hilbert's philosophy was mainly concerned with mathematics proper. On the other hand, it was in economics that Wald achieved the existence proof. In this respect, the Vienna Circle played a key role in connecting the philosophy advocated by Hilbert with all sciences in general. In the Circle's view, all sciences must be pursued according to the same method. The Vienna Circle considered that the axiomatic method is indispensable for this program and took a formalistic approach to an axiomatic system. In considering the emergence of Wald's existence proof in economics, we have to distinguish the influence of Hilbert's philosophy from that of the Vienna Circle's philosophy. The former had an influence on Wald's concern for the existence proof, while the latter had an influence on his concern for the axiomatic treatment of economic theory. If we ambiguously grasp the content of Hilbert's formalism, we fail to make a clear distinction between the influence of Hilbert's philosophy and that of the Vienna Circle's philosophy of science.

Then we define the content of Hilbert's formalism in section 3. And we show how the Viennese mathematicians accepted Hilbert's formalism.

In section 4, we deal with the new trend in the philosophy of science in Vienna and study its relation with the axiomatic economic analysis. And we examine the view of unified science promoted by the Vienna Circle stimulated Wald to axiomatize economic theory.

In section 5, we examine the mathematical structure of Wald's proof. Wald achieved his existence proof in an elementary way. We are inclined to have an impression that Wald's proof had not contained mathematical contents attracting the Viennese mathematicians' interest.

In Vienna, the mathematical tools of convex analysis were intensively studied. Then we examine to what extent Wald's proof linked up with the development of convex analysis in Vienna at that time.

Wald's model contains the system of linear inequalities. The theory of linear inequalities is closely related to the Minkowski-Farkas lemma, which is often referred to as the separation theorem of convex sets. Although Wald did not mention the Minkowski-Farkas lemma, we can prove it based upon his idea of proof. We may conclude that his idea of proof contained substantially the mathematical contents of the same depth as the Minkowski-Farkas lemma.

In addition, based upon his idea of proof, we can prove the duality theorem in linear programming and the minimax theorem. These mathematical exercises may help us to evaluate the "depth" of Wald's proof. The proofs are collected in the Appendix.

Finally, the summary of this paper is provided in the concluding remarks.

2. Viennese view on Cassel's system

It was the Walras-Cassel equation system that Wald gave an existence proof. We first present Cassel's formulation. In his "Grundriss einer elementaren Preislehre" [10] and in his *Theoretische Sozialökonomie* [11], Cassel reformulated Walras' general equilibrium theory as the following equation system.

Suppose that there are n goods and m factors of production. The available quantities of factors are assumed to be given exogenously. We denote the demand for j-th good by s_j and the supply of i-th factor by r_i respectively, where $j = 1, 2, \cdots, n$ and $i = 1, 2, \cdots, m$. Let p_j represent the price of j-th good and q_i the price of i-th factor respectively. p and q represent the column price vectors with component p_j and q_i respectively.

Let a_{ij} be the amount of i-th factor required for the production of one unit of j-th good. We define A as the m by n matrix with elements a_{ij}. We denote by r an m-dimensional column vector with components r_i and by s an n-dimensional column vector with components s_j. Cassel's system can be summarized as follows.

(1) the demand-supply conditions for factors:

$$As = r. \tag{1}$$

(2) the price-cost conditions for goods[3]:

$$^{t}Aq = p. \tag{2}$$

(3) the demand functions for goods:

$$s_j = f_j(p_1, p_2, \cdots, p_n) \quad (j = 1, 2, \cdots, n). \tag{3}$$

Karl Menger introduced this theory in his colloquium and frequently discussed it with the members. During the interwar period, the research activities gradually shifted from the university to the private seminar and the research institute (see E. Craver [13]).

In 1920s, Hans Mayer's seminar focused on the *Zureschnungsproblem* (the imputation problem), the problem of imputing the values of consumption goods to the factors of production. Karl Menger, Schlesinger, Wald and Morgenstern attended the Mayer's seminar.

In the early thirties, Ludwig von Mises' seminar played an important role in the Viennese economics community. Among other economists who attended

---

[3] Here ^{t}A stands for the transposed matrix of A.

the Mises' seminar, we must mention such names as Friedrich von Hayek, Fritz Machlup and Gottfried Haberler. Karl Schlesinger was listed as one of the regular members of the seminar (see L. von Mises [48], p.100). The range of their discussions in the seminar was extended beyond economics. In his *Notes and Recollections*, Mises ([48], p.97) wrote; "we informally discussed all important problems of economics, social philosophy, sociology, logic, and the epistemology of the sciences of human action".

It would be plausible to infer that through the debate in the Mayer's seminar, Menger, Schlesinger and eventually Wald became to be familiar with the Walras-Cassel equation system.

Certainly, Cassel's *Theoretische Sozialökonomie* was used as a textbook so widely in Central European countries (see E. R. Weintraub [57], p.4) that we can find several criticisms on this treatise.

In 1926, Morgenstern wrote a review article on Cassel's book, but it was never published. In addition, his article in the *Encyclopaedia of the Social Sciences*, Morgenstern ([34], p.367) argued that "Cassel took over Walras' equations in a simplified form, but in his presentation there are more equations than unknowns; that is, the conditions of equilibrium are overdetermined[4]".

Edward Schams [42], who kept a close relationship with Morgenstern, argued that one of Cassel's equations was not independent. And Schams [42] condemned Cassel for his mathematical mistake[5].

---

[4] Regrettably, we have no idea about what Morgenstern intended to mean by this passage. It seems to be a conundrum.

[5] Although he made no reference to Wicksell's 1900 paper [58], it seems that Schams' comment was based upon his reading of the paper. Wicksell was the first who refuted the Cassel's equation system as logically inconsistent. In addition, Wicksell was the first who pointed out that the equality of the demand-supply conditions for factors (1) should be replaced by inequality. Wicksell [59] pointed out that Cassel's equation system had more unknowns than equations since the system had to fulfil the Walras' law. Comparing with Cassel, Walras showed a correct way to equate the number of unknowns to that of equations by introducing the numéraire. In 1919, in his critical review of Cassel's *Theoretische Sozialökonomie*, Wicksell [59] noted that Cassel proceeded in an entirely different manner. Cassel considered money had a function of a store of value, but Wicksell insisted that Cassel's "premature introduction of money had contributed not to increased lucidity but rather the reverse (p.225)". In his 1919 paper, we read;

> he (Cassel) describes how the total rewards of the factors of production are in the main identical with total (real) incomes, and are at the same time the source of the demand for goods and services; he adds that these incomes are not all consumed, but are partly saved. But at this point the equality between the sum of the factors of production now available, and that part of them which enters into the various goods demanded for consumption, ceases to obtain, and Professor Cassel's equations (in this paper (1)) is no longer valid. (pp.225-226)

Thus we can safely say that Cassel's system was known to the Viennese economists by the mid-1920s.

In the thirties, some authors, such as Neisser, von Stackelberg and Zeuthen, pointed out that the equality of the number of equations and that of unknowns did not assure an economically meaningful solution. Neisser ([36], p.424) gave a numerical example in which equilibrium prices are negative. Von Stackelberg ([51], pp.463-467) noted that if the number of factors exceeded that of goods, there were more equations than unknowns. Zeuthen [61] argued if the demand for any factor fell short of its supply, then the factor could no longer be regarded as a scarce good but should be regarded as a free good. Then he insisted that the equation (1) should be modified to the inequality and that he introduced the idea of complementary slackness conditions.

Karl Schlesinger, independently of Zeuthen, made the same point[6]. Karl Menger invited him to present the paper in Menger's colloquium.

In this way, the Walras-Cassel system was widely discussed in Vienna. Karl Menger introduced this theory into his colloquium. Menger seemed to regard the Walras-Cassel general equilibrium theory as a suitable material of the discussion in his colloquium.

However, the foundation of general equilibrium theory was criticized by the Central European economists. Mayer, von Mises, A. Amonn and E. Schams seemed to be the representative figures. They criticized general equilibrium theory since it could not explain the principles of price formation in real markets.

Mises ([48], p.36) claims that "the Austrian School endeavors to explain prices that are really paid in the market, and not just prices that would be paid under certain, never realizable conditions". The Austrian school rejects a mere description of a state of hypothetical static equilibrium. They insist that the economists should describe and explain the entire genetic path of the economic process in its various intermediate stages leading to the attainment of the equilibrium. The origin of this Austrian attitude goes back to Carl Menger's methodology of economics.

Amonn [3], Schams [42] and Mayer [27] criticized Cassel's economics as a whole on the same grounds. Cassel began directly with a demand function. They criticized that Cassel's theory was devoid of the explanation of price formation ([3] p.41, [42] pp.389-390, [27] pp.234-235).

Karl Menger, the son of Carl Menger, was fully aware of these criticisms. Nevertheless, the members around Menger adopted the Cassel equation system

[6] Schlesinger read the papers by Neisser and von Stackelberg. Thanks to their papers, Schlesinger obtained the idea of modifying the equation (1) to the inequality. A footnote indicated that Schlesinger became aware of Zeuthen's paper while his paper was in proof. See K. Schlesinger [44].

as a material of the discussion in the colloquium. We explore the reason why they accepted the equilibrium theory.

Karl Menger ([31], p.54) explains the methodological difference between the Austrian economists and the mathematical economists; the former look for *causal explanations*[7] of economic phenomena, whereas the latter wish to confine themselves to the study of *functional relations* (K. Menger ([31], p.54). Moreover, the Austrians are looking for the ultimate cause of economic phenomena.

On the contrary, the mathematical economists refuse the causal explanations of economic phenomena since we are easily led to the metaphysical or pseudo-problems by such a research program. Menger ([31], pp.54-55) goes on to say; "the Austrians are looking for the *essence* (*das Wesen*) of economic phenomena, thereby moving on dangerous ground surrounded by swamps of pseudo-problems". The members of the colloquium consciously avoided the danger of stepping into the area of metaphysics or pseudo-problems. For example, in his review of Otto Kühne's *Exakte Nationalökonomie*, Franz Alt criticizes that Kühne adopts the causal explanations of economic phenomena. Alt [2] claims that the causal explanations will be a source of metaphysical arguments beyond our observations. Alt insists that economics should be restricted to the research of economic phenomena which can be expressed in terms of observable languages.

According to Karl Menger ([31], p.47), one of the reasons why he adopted Cassel's system as a material of the discussion in his colloquium was that all objects in the system were clearly measurable or observable[8]. The members of the colloquium rejected the causal explanation of economic phenomena. They intended to exclude the metaphysical problem from an economic theory.

In this respect, the Walras-Cassel system seemed to be a suitable material. It was free from such a metaphysical problem since it dealt with the functional explanation among the economic quantities.

---

[7] Here we make clear the meaning of *causal explanation*. What do we mean when we say, "A is the cause of B"? In the opinion of philosophers of science, causal relation means predictability. This does not mean *actual* predictability, rather a *potential* predictability. It means predictability in the sense that, if all the relevant facts that surrounded A were known, together with all the relevant laws, the occurrence of B could then be predicted. In other words, we say that an event B is caused by a preceding event A if and only if B is deducible from A with the aid of the relevant laws. See R. Carnap [9], chap.19.

[8] Cassel claims that the concept "utility" should be excluded from economic theory. He considers that the concept "utility" is nonobservable term in the sense that it can never be measured by any simple and direct procedure. We do not have a clear idea whether the members of the colloquium admitted or rejected the concept of utility. In 1936, Alt published the paper on the measurability of utility [1]. Presumably, there were different attitudes among the members of the colloquium toward Cassel's suggestion.

In this way, the members of the colloquium adopted the Walras-Cassel system.

They were exclusively concerned with the existence problem of the Walras-Cassel system. In this respect, they were strongly influenced by the philosophy advocated by Hilbert. We discuss this point in detail in the subsequent section.

3. Hilbert's philosophy on foundations of mathematics

Wald's primary concern was the existence proof of the Walras-Cassel system. In our view, Wald was strongly influenced by the philosophy advocated by Hilbert. We first define the content of Hilbert's formalism.

Hilbert required the axioms to satisfy certain logical requirements, i.e., completeness, independence and consistency. Among them, the check of the last requirement is the most difficult task. Axioms are said to be consistent when it is not contradictory, i.e., when it is not the case that both a formula and its negation are provable in the system.

Moreover, Hilbert insists that the mathematical existence of the concept is established if one can prove that the attributes assigned to its concept never lead to a contradiction. Hilbert ([19], p.300) says;

> If contradictory attributes be assigned to a concept, I say that *mathematically the concept does not exist.*

According to Hilbert, the proof of the consistency is equivalent to the existence of the corresponding mathematical concept[9] . We regard this insistence as the most essential feature of Hilbert's formalism.

Hilbert's formalism contrasts sharply with the intuitionism developed by L. E. J. Brouwer. Menger intended to study with Brouwer, who was regarded as one of the leading figures in the field of topology at that time. In 1925, Menger went to Amsterdam, where he spent two years working with Brouwer. But, in the twenties, Brouwer's major interest had shifted to the foundations on mathematics. After his return to Vienna, Menger introduced Brouwer's intuitionism into his colloquium and frequently discussed it with the members.

Brouwer's view on the foundations of mathematics consists of two claims.

---

[9] Historically speaking, we can find an axiomatic method in Euclid's *Elements* in the third century B.C.. Hilbert's view is much more penetrating and solid than Euclid's. In this respect, Bourbaki ([7], p.39) said;

> whereas in traditional logic the non-contradiction of a concept only made it "possible", it was equivalent for Hilbert (at least for mathematical concepts defined axiomatically) to the *existence* of the concept.

(1) Brouwer believes that mathematics is founded on intuition[10].
(2) Brouwer rejects the law of excluded middle applied at least to infinite sets[11]. He accepts only proof that is constructive. Brouwer considers that a mathematical object exists only when we can actually construct the object in a finite number of steps.

If Brouwer's view were to be adopted, a great part of mathematics would have to be given up[12]. In his *Reminiscences*, Menger ([33], pp.138-139) said that "his attacks on the law of the excluded middle and the consequences for mathematics of its rejection had been discussed in the Circle on several earlier occasions", but Brouwer's "obscure remarks on primordial intellectual phenomena and primordial mathematical intuition were not taken seriously by any member of the Circle".

Hilbert's philosophy had a great influence on mathematicians' world. Since it appeared, mathematicians recognized the significance of the existence proof. They considered that the consistency of a mathematical theory with which they were concerned should be proved prior to any mathematical reasoning.

In the field of the foundations of mathematics, Hilbert intended to prove that all of mathematics contained no contradiction.

In this connection, a young Viennese mathematician, Kurt Gödel made a completely unexpected and most significant discovery. Roughly speaking, Gödel's incompleteness theorem states that for any formal theory there exists a proposition which can neither be proved nor disproved within the theory in question. Gödel's result shattered Hilbert's hope for founding mathematics on proof of its consistency[13].

The members of the colloquium were quite interested in the foundations of mathematics and encouraged the research on the subject. Hans Hahn was an advisor of Gödel's dissertation thesis on completeness theorem. It was at the

---

[10] According to Hahn, "its point of departure seems too much akin to Kant's pure intuitionism and Kant's *a priori*". H. Hahn, [17] p.26.

[11] Scarf ([41], p.12) insisted that "Brouwer's original demonstration in 1910 was not concerned with effective computational procedures, and Brouwer himself eventually rejected the theorem because of its "non constructive" aspects". Scarf's insistence seems to be based upon the fact that Brouwer published the paper [8] concerning to the fixed point theorem, which was modified so as to meet the intuitionistic standards.

[12] For example, Brouwer does not accept the general concept of irrational number since we can not construct its object in a finite number of steps. On the contrary, Hilbert insists that in order to prove the existence of irrational number, we do not actually have to construct it and we do not even have to show how it can be constructed. All we have to do is to prove that it must exist because any other conclusion results in a contradiction.

[13] It is well-known that von Neumann pursued his interest in problems concerning the foundations of mathematics. For instance, his paper [49] is devoted to the problem of consistency of mathematics.

mathematical colloquium that Gödel presented his result on the incompleteness theorem [15]. At the time Menger was on a lecture tour in the United States. Menger ([33], p.171) said; "I learned of his epoch making logical discovery from a letter of Nöbeling while I was lecturing in Houston". And he ([33], p.202-203) wrote; "In my excitement about this news I interrupted my course with a report about Gödel's epoch-making discovery".

Although he published no paper on the foundations of mathematics, Wald seemed to be decisively influenced by Hilbert's philosophy.

Then we raise here the question; how did Hilbert's thought exert an influence on Wald's research? In order to give an answer to this question, we restrict here ourselves to the field of geometry, with which Wald's early research was connected.

In his *Grundlagen der Geometrie* (*The Foundation of Geometry*), published in 1899, Hilbert established a complete set of independent axioms by means of which it would be possible to prove all theorems of Euclidean geometry. And by applying his axiomatic method, Hilbert showed that any contradiction in Euclidean geometry must appear as a contradiction in the arithmetic.

Wald's early research on geometry was concerned with the axiomatic system of Hilbert's *Grundlagen*. Wald [52] improved Hilbert's system by omitting some axioms and weakening others.

The *Ergebnisse* contains no paper on the axiomatic system of Hilbert's *Grundlagen* except for Wald's paper. It should be noted that it was not in Vienna but in Göttingen that Hilbert's *Grundlagen* was intensively studied. Hilbert became a professor at the University of Göttingen in 1895. At that time Göttingen was one of the centers of mathematical research activities. Among many young mathematicians attracted to Göttingen from all over the world by Hilbert's *Grundlagen*, we can count the big names as F. Schur and A. Rosenthal. They studied Hilbert's book and some of their results were devoted to the improvements of its axiomatic system. A great part of these papers on the foundations of geometry appeared in the *Mathematische Annalen*. Hilbert was one of the principal editors of the journal. We recall here that the movement toward an axiomatization in economics never occurred in Göttingen, Hilbert's homeground.

On the other hand, the Viennese mathematicians and philosophers resonated with the philosophical aspect of Hilbert's *Grundlagen*. Indeed, the Vienna Circle paid a special attention to Hilbert's *Grundlagen* in order to support their philosophical views. And this fact seemed to play an important role in the emergence of highly axiomatic mathematization of economics in Vienna. We examine this point in detail.

4. The new trend in the philosophy of science

The Vienna Circle was formed around the physicist and philosopher, Moritz Schlick. All members had a first-hand acquaintance with some field of science, either mathematics, physics or social science. Also they were familiar with logic and keenly interested in foundational questions. The central figures among them were Rudorf Carnap, Otto Neurath, Hahn and many others. They met weekly since 1924. It was in the fall of 1927 that Karl Menger began to take part in the Circle on a regular basis.

Their essential views might be classified under the following aspects[14].

(a) They sharply reject the existence of the synthetic *a priori* propositions in the sense of Immanuel Kant[15].

---

[14] Their views on the role of philosophy were consonant with Wittgenstein's views, which were manifested by the closing statements in his *Tractatus Logico-Philosophicus*. The members of the Vienna Circle studied the *Tractatus* at Schlick's and Hahn's request. In the *Tractatus*, we read;

> The right method of philosophy would be this. To say nothing except what can be said, *i.e.* the propositions of natural science, *i.e.* something that has nothing to do with philosophy: and then always, when someone else wished to say something metaphysical, to demonstrate to him that he had given no meaning to certain signs in his propositions. This method would be unsatisfying to the other—he would not have the feeling that we were teaching him philosophy— but it would be the only strictly correct method. ······ Whereof one cannot speak, thereof one must be silent. ([60], pp.187-189).

In Karl Menger's opinion ([33], p.84), "this last sentence of the *Tractatus* was of great importance for philosophy". But the Vienna Circle's philosophy did not directly originate in Wittgenstein's *Tractatus*. Karl Menger ([33], p.90) wrote;

> Schlick emphasized this role of philosophy in his lectures even before he had seen the *Tractatus*". This concept was probably also dormant in the minds of some of the other members of the Circle".

For detailed accounts of the relation between Wittgenstein and the members of the Vienna Circle, see Menger [33].

[15] In all subject-predicate judgements, Kant ([22], p.130) argued that the relation of a subject A to the predicate B could be considered in two different ways. The relation is either the predicate B belongs to the subject A as something that is contained in the concept A, or B lies entirely outside the concept A. In the first case Kant calls the judgement analytic, in the second case he calls the judgement synthetic. In this sense, an analytic statement involves nothing more than the meaning relations of the term, while a synthetic statement is an assertion that goes beyond the assigned meanings of the term. A priori statements are completely independent of experience. It is never necessary to refer to experience as a justification for the truth of an analytic statement. A posteriori statements are essentially dependent on experience in the sense that it has to be justified by experience. It is obvious that all analytic statements are a priori. Now an important question arises; does the demarcation line between a priori and a posteriori coincide with the one between analytic and

(b) They consider that the task of philosophy is to investigate the syntactical structure of propositions in sciences.
(c) The peculiar view they put forward is that all sciences must be pursued according to the same method. In this sense, there can exist only one science for them. They were actually fighting for *unified science.*

They stressed the need for rigorous transformation of propositions through logico-deductive inference and critically examined the statements in sciences from a formal point of view. In order to approach this target, the axiomatic method seems to be indispensable. In an axiomatic method, all the assumptions required for the given theory have to be enumerated explicitly and completely. In addition, the conclusions have to be drawn by the logico-mathematical reasoning.

Furthermore, according to the program for unified science, there is no fundamental difference in axiomatic method in all sciences.

The Vienna Circle takes a formalistic approach, which is ascribed to Hilbert, to an axiomatic system. Carnap says; "Today, we often take a purely formalistic approach to an axiom system. We do not ask what interpretations or applications it may have, but only whether the system of axioms is logically consistent and whether a certain statement is derivable from it"[16].

The Viennese mathematicians and economists were in favor of the Vienna Circle's philosophy of science. In their acceptance of it, they each put a special emphasis on one of aspects according to each interest.

Morgenstern ([35], p.396) stressed the need to treat any theories in a rigorous way and was in favor of the axiomatic method. He said that "in order to gain a rigorous insight into the state of any science, the use of the axiomatic method cannot be dispensed with". And he continued to say, "there is no fundamental difference in the axiomatic procedure whether it be the formulation of empirical or aprioristic sciences ([35], p.396)." For economics, "this science has to be seen in its empirical character and has to be built as such, but that it has to be developed as rigorous theory ([35], p.396)". It is obvious that Morgenstern was strongly influenced by the view of unified science promoted by the Vienna Circle.

Morgenstern grasped the content of axiomatic method as rigorous logico-deductive system. Although he ([35], pp.409-410) appreciated the Wald's exis-

---

synthetic? In other words, is it possible for knowledge to be both synthetic and a priori? Kant answered to this question affirmatively. Kant ([22], chap.1) regarded all geometrical statements as synthetic a priori propositions. In the Circle's view, Kant should be condemned by his confusion between *pure* geometry and *physical* geometry. The former is simply a deductive system based on the set of axioms. It is completely independent of the world. The latter, on the other hand, is concerned with the application of pure geometry to the world. This distinction was made especially clear through Hilbert's *Grundlagen.*

[16] R. Carnap. [9] p.130.

tence proofs, Morgenstern were rather concerned with the application of economic theory to the real world and the empirical justification for economic theory.

On the other hand, as a mathematician, Menger thought he should be concerned with the rigorous analysis of theories. His methodological views are stated in his paper [28] on the law of returns. Although he admits that "the crucial issue for economics to be whether or not these laws are empirically confirmable"(p.280), Menger regards this empirical question as a secondary issue[17].

Both Wald and Alt shared Menger's methodological views in common (see A. Wald [55] p.369., see also F. Alt [2]).

Wald was not a member of the Vienna Circle. However, Wald seemed to become acquainted with the Vienna Circle's view through his reading of Menger's papers or personal communication with Menger.

Wald axiomatized economic theory. The Vienna Circle's philosophical view had an influence on Wald's axiomatic treatment of economic theory. For him, axiomatizing the theory was fundamental prerequisite for the existence proof. Hilbert's formalism had an influence on Wald's concern for the existence proof. As a mathematician, Wald studied Hilbert's *Grundlagen* and its axiomatic method from purely mathematical interest. In consequence, the emergence of an axiomatization in economics due to Wald is explained by his acceptance of Hilbert's formalism in connection with the Vienna Circle's philosophical views.

5. The development of convex analysis

We recall here that it was at the mathematical colloquium that Wald presented his proof and its main participants were mathematicians. If Wald's proof had not contained mathematical contents attracting the Viennese mathematicians' interest, Wald would never have been induced to present his proof in the colloquium.

In Vienna, the mathematical tools of convex analysis were intensively studied. For example, Eduard Helly at the University of Vienna established a theorem which ensures the existence of an intersection of a family of convex sets. It greatly contributed to a linear inequality theory. Hahn established what is now called the Hahn-Banach theorem. From a geometric viewpoint, it asserts that there exists a closed hyperplane that separates two disjoint convex sets in infinite dimensional space. Moreover, the fixed point theorem was intensively

[17] Karl Menger had a skeptical view as to the possibility of unified science. He ([33], p.176) said, "I feared that the idea of a unified science might possibly lead to the exclusion *a priori* of potentially valuable objects or methods of study."

studied by Borsuk, Knaster and Mazurkiewicz. They were frequently invited to the colloquium.

In the case of von Neumann's paper[18], we can easily recognize that his proof closely linked up with the development of convex analysis. In fact, he formulated a balanced growth model in terms of inequalities and his existence proof was based upon a fixed point theorem.

On the contrary, at first glance we have an impression that Wald's proof, being based upon the mathematical induction, had nothing to do with the development of convex analysis of that time. However, it should be noted that it was Karl Menger who encouraged him to present his proof at the mathematical colloquium. It may be conjectured that Wald's idea of proof also linked up with the development of convex analysis.

Then we ask the question; to what extent did Wald's existence proof rest on the properties of convexity? In order to give an answer to this question, we examine Wald's proof and evaluate its mathematical structure.

As we have already mentioned above, it was Karl Schlesinger who started the discussion. The model which Schlesinger reformulated may be summarized as follows.

(4) the demand-supply conditions for factors:

$$As \leqq r. \tag{4}$$

(5) the price-cost conditions for goods:

$$^{t}Aq = p. \tag{5}$$

(6) the demand functions:

$$s_j = f_j(p_j) \quad (j = 1, 2, \cdots, n). \tag{6}$$

(7) the complementary slackness conditions:

$$^{t}q(r - As) = 0. \tag{7}$$

Schlesinger also raised the conjecture that this procedure would solve the existence problem.

Wald inverted the demand function (6) to the form

$$p_j = f_j(s_j) \quad (j = 1, 2, \cdots, n). \tag{8}$$

[18] This paper was presented for the first time in 1932 at the mathematical seminar of Princeton University. Karl Menger invited him to present the paper at the mathematical colloquium.

Wald proved the theorem that the system of equations (4)(5)(7) and (8) has a solution under the following conditions (see Appendix (I)).

(A1) $A \geqq 0$, and for each j there is at least one i for which $a_{ij} \neq 0$.
(A2) $r > 0$.
(A3) For each j, the function $f_j(s_j)$ is defined for every positive value of s_j, its value is nonnegative, continuous , and strictly monotone decreasing, and in addition $\lim_{s_j \to 0} f_j(s_j) = +\infty$.

The idea of Wald's proof was based on the mathematical induction by the number of goods.

It can be readily seen that the system of equations (4)(5)(7) and (8) has a solution for the case $n = 1$. It now suffices to prove the theorem for the case n on the inductive hypothesis that it holds for the case $n - 1$.

We define $\overline{\lambda} = \max\{\lambda > 0 \mid r_i - a_{in}\lambda \geqq 0, \;\; i = 1, 2, \cdots, m\}$. For each $\lambda \in (0, \overline{\lambda})$, consider the system, which we denote by $(\mathrm{W}_{n-1}^{\lambda})$,

$$(\mathrm{W}_{n-1}^{\lambda}) \begin{cases} a_{i1}s_1 + \cdots + a_{i,n-1}s_{n-1} \leqq r_i - a_{in}\lambda = r_i' \quad (i = 1, 2, \cdots, m), \\ p_j = f_j(s_j), \\ p_j = a_{1j}q_1 + \cdots + a_{mj}q_m \quad (j = 1, 2, \cdots, n-1), \\ q_i(a_{i1}s_1 + \cdots + a_{i,n-1}s_{n-1} - r_i') = 0. \end{cases}$$

For $\lambda \in (0, \overline{\lambda})$, the system $(\mathrm{W}_{n-1}^{\lambda})$ satisfies assumptions (A1) to (A3). Then the system $(\mathrm{W}_{n-1}^{\lambda})$ has solutions $s_j^* > 0$, $p_j^* \geqq 0$ $(j = 1, \cdots, n-1)$ and $q_i^* \geqq 0$ $(i = 1, \cdots, m)$ by the inductive hypothesis. We choose $s_n = \lambda \in (0, \overline{\lambda})$[19]. The solutions for the case $n - 1$ and $s_n = \lambda$ satisfy all the equations except for those that contain the unknown p_n, that is to say,

$$p_n = f_n(\lambda) \;\text{ and }\; p_n = a_{1n}q_1 + \cdots + a_{mn}q_m.$$

Then we construct the set,

$$T(\lambda) = \{a_{1n}q_1 + \cdots + a_{mn}q_m - f_n(\lambda) \mid (q_1, \cdots, q_m) \text{ are solutions of } (\mathrm{W}_{n-1}^{\lambda})\}.$$

The essential argument of Wald's proof is to show that there exists $\lambda_0 \in (0, \overline{\lambda})$ such that $0 \in T(\lambda_0)$.

To obtain this result, Wald applies the intermediate value theorem to the correspondence $T : (0, \overline{\lambda}) \twoheadrightarrow \mathbb{R}$. In order to appeal to this theorem, we have to check a couple of points; that is to say,

(i) T is a convex-valued and bounded-valued correspondence that has a closed graph on any closed subinterval I of $(0, \overline{\lambda})$, and

[19] We are much indebted to Hildenbrand [20] for an important insight into the mathematical ideas embodied in Wald's proof.

(ii) there are λ' and λ'' in I such that $T(\lambda')$ contains a non-positive number and $T(\lambda'')$ contains a nonnegative number.

The remaining part of Wald's proof ([53], pp.15-16) is devoted to showing the fulfillment of all the conditions of the intermediate value theorem. Thus, by virtue of this theorem, there exists $\lambda_0 \in (0, \overline{\lambda})$ such that $0 \in T(\lambda_0)$. Then $s_n = \lambda_0$ together with the solution of $(\mathrm{W}_{n-1}^{\lambda_0})$ constitute the solution of the equation system (4)(5)(7) and (8).

We make here a few remarks on the characteristics of Wald's original method of proof.

(I) Wald assumed that supply of the factor is exogenously given and does not depend on the factor price. This assumption enabled Wald to solve the existence problem in an elementary way. The first step is to solve the system of inequality (4), which can be reduced to the problem that Julius Farkas proposed. Assume now that the desired solution s^* actually exists. Then, one can obtain from (8) the solution p^* corresponding to this s^*. Finally, we determine the value q^*, which must be chosen so as to satisfy both conditions (5) and (7).

The model developed by Schlesinger and Wald is based on the idea that the system of equations should be modified to that of inequalities. Schlesinger introduced nonnegative slack variables on the demand side of the factor market equations. The condition (7) implies that if the condition (4) holds with strict inequality for some factor i, then the corresponding price q_i must be zero.

(II) As mentioned above, Wald's idea of proof was based on the mathematical induction by the number of goods. His proof was not based on the fixed point theorem. Indeed, the proof of the existence for modified Walras-Cassel system could be achieved in much more elementary ways without the fixed point theorem.

If we assume that the supply of factor does depend on the factor prices, the fixed point theorem must be needed for proving existence. Indeed let us assume the supply of factor r_i is a function of its factor price q_i. We start with a vector $q' = (q'_1, q'_2, \cdots, q'_m)$, which is chosen arbitrarily. We solve the system of inequality (4) for the vector q' and we obtain the solution s'. From (8) the vector $p' = (p'_1, p'_2, \cdots, p'_n)$ is obtained for the solution s'. Finally, we determine the value q'', which must be chosen so as to satisfy both condition (5) and (7). However, there is no guarantee that the vector q'' coincides with the initial vector q'.

Arrow [4] introduces a Wald's episode in connection with the use of fixed point theorem. In the Wald's paper, which was announced to appear in the *Ergebnisse* in 1938, he formulated a general equilibrium theory in a pure exchange economy on the basis of Paretian analysis. Wald ([55], pp.379-380) referred repeatedly to the need for "method of modern mathematics" for the proof. Unfortunately, Wald's paper was lost. Taking Uzawa [47]'s result into account, the existence of equilibrium is mathematically equivalent to the fixed point theorem. Then, as Arrow ([4], p.26) pointed out, "it seems impossible to conceive of any proof not based on a fixed point theorem". Thus Arrow ([4], p.26) conjectures that "the references to "modern methods" may imply Wald's use of fixed point arguments". Arrow ([4], p.27) goes on to say; "But we will never know, unless, against all probability, Wald's original manuscript emerges".

It is often said that Wald's proof is "clumsy" (for example, see Wolfowitz, ([56], p.3)). Thus, for example, Kuhn [24] reorganized Wald's proof by using the duality theorem in linear programming and the Kakutani fixed point theorem. In this paper, we do not take such an approach. Rather, we are going to examine to what extent mathematical theorems can be proved based upon Wald's idea of proof. For example, we can prove the Minkowski-Farkas lemma with the additional conditions (A1) and (A2) (see Appendix (II)).

Moreover, as is well known, the duality theorem in linear programming can be proved by using the Minkowski-Farkas lemma. We assume the conditions (A1) and (A2') $r \geqq 0$ and $p \geqq 0$. We replace the equality of the price-cost conditions for goods (5) by the inequality

$$^{t}Aq \geqq p \tag{9}$$

and introduce the complementary slackness conditions

$$^{t}s(^{t}Aq - p) = 0. \tag{10}$$

It can be verified that Wald's idea is applicable to the proof of the duality theorem in linear programming with these additional conditions (see Appendix (III)). In addition, we consider the two-person zero-sum matrix game and the minimax theorem. Wald's idea is applicable to the proof of the minimax theorem (see Appendix (IV)).

Thanks to these mathematical exercises, we can evaluate the "depth" of Wald's proof. The result of these exercises shows that Wald's proof contained mathematical contents attracting Viennese mathematicians' interest. Then Menger seemed to consider that Wald's paper was well-qualified as a material of the discussion in a seminar of mathematicians.

6. Concluding remarks

As we have mentioned above, Wald's existence proof should be regarded as a starting point of the axiomatization in economics. In this paper, we have investigated the intellectual milieu which stimulated Wald to the existence proof; (a) the research activities in economics, (b) Hilbert's philosophy on the foundations of mathematics, (c) the new trend in the philosophy of science, and (d) the development of convex analysis. It was the interplay of these factors that stimulated Wald to the existence proof. Karl Menger played an important role as an organizer. He built the bridge between these factors.

We are now in a position to give a brief summary of this paper in comparison with the preceding literature on this topic.

In section 2, we explored the reason why the members of Menger's colloquium accepted the Walras-Cassel general equilibrium theory as a material of the discussion in the seminar. Weintraub [57] omitted the question which we posed in this paper. According to Weintraub, Cassel's book was widely used as a text in Central European countries. However, we must pay attention to the fact that the general equilibrium theory was severely attacked from the Austrian economists such as von Mises and Mayer. We investigated the methodological views shared by almost all members in Menger's colloquium. They rejected the causal explanation of economic phenomena and looked for the study of functional relations. In addition, Cassel's equation system was suitable for a material of the discussion since all objects in the system were observable.

Wald axiomatized the Walras-Cassel system and solved the existence problem.

In our views, Wald's concern for the existence proof reflected Hilbert's philosophy on the foundations of mathematics. And his concern for the axiomatic treatment in economics reflected the Vienna Circle's philosophy. We have to distinguish the influence of Hilbert's philosophy from that of the Vienna Circle's philosophy.

Then, in section 3, we defined the content of Hilbert's formalism as following; the consistency of the axiomatic system is equivalent to the existence of the mathematical concept.

In contrast, Ingrao & Israel ([21]) and Punzo ([37], [38], [39]) seemed to regard the essential content of Hilbert's formalism as purely logico-deductive systems (see L. Punzo. [39] p.23).

For example, Ingrao & Israel ([21], pp.182-183) said that Hilbert's "*Grundlagen der Geometrie* represented the programmatic manifesto of the axiomatic movement and explicitly brings out the main points of this tendency, according to which a mathematical theory is nothing more than a complex of theorems obtained through deductive logic and defining the properties of a mathematical entity defined by axioms".

Ingrao & Israel ([21], pp.182-183) seemed to regard that the influence of Hilbert's philosophy was reflected only in Wald's attitude toward axiomatizing economics. This is due to their ambiguous grasp of the content of Hilbert's formalism.

Hilbert's philosophy was mainly concerned with mathematics proper. On the other hand, it was in economics that Wald achieved the existence proof. On this point, Ingrao & Israel ([21], p.187-188) pointed out that Wald was strongly influenced by the ideas of the Vienna Circle. They ([21], p.188) said that Vienna was "the center of renewed interest in the formalization of the social sciences". Ingrao & Israel ([21], p.189) regarded the fundamental theme of the Vienna Circle's philosophy as the "unification of the sciences".

Following the Vienna Circle's view, all sciences must be pursued according to the same method. In order to approach this target, the axiomatic method seems to be indispensable.

In our views, the emergence of the existence proof in economics due to Wald can be explained by his acceptance of Hilbert's formalism in connection with the Vienna Circle's philosophy of science.

Finally, we have evaluated the mathematical depth of Wald's existence proof. This aspect has never been considered in the preceding investigations in the recent literature. We have shown that the important theorems in the field of convex analysis such as the Minkowski-Farkas lemma, the duality theorem in linear programming and the minimax theorem can be proved based upon Wald's idea of proof.

Mathematical economics flourished in Vienna between the Wars, which seemed to forecast a brilliant future. However, it could not be kept for long. Hitler's annexation of Austria ended that possibility.

Politically, Austria was in a state of increasing tension and chaos. Nazi menace became urgent. Menger ([30], p.19) said;

> The *Ergebnisse* was criticized (with specific reference to Wald) for its large number of Jewish contributions just when I felt that we ought to honor that journal by making Wald co-editor. Issue 7 was edited by Gödel, Wald and myself. But Issue 8 containing Wald's paper on collectives was destined to be the last of the series. ··· Viennese culture resembled a bed of delicate flowers to which its owner refused soil and light while a fiendish neighbor was waiting for a chance to ruin the entire garden.

The activities in the Vienna Circle and Menger's colloquium came to an end. The members were dispersed.

Menger left for the United States and was offered a professorship at the University of Notre Dame in Indiana. Carnap also emigrated to the United States in 1935, where he taught at the University of Chicago. In 1936, Schlick was assassinated by a student on the stairways of the University of Vienna. Karl Schlesinger committed a suicide exactly on the day of Hitler's invasion to Austria.

Wald was reluctant to leave Vienna. But Morgenstern persuaded Wald to leave Austria and got him to the United States.

7. Appendix

In this Appendix, we evaluate the "depth" of Wald's proof. As we have already said in section 5, we can prove the Minkowski-Farkas lemma, the duality theorem in linear programming and the minimax theorem based upon Wald's idea of proof.

At first, we briefly repeat Wald's original method of proof for the convenience of our arguments. Next, we shall show that Wald's idea of proof is applicable to the proof of the Minkowski-Farkas lemma, the duality theorem in linear programming and the minimax theorem. These mathematical exercises may help us to evaluate the "depth" of Wald's idea.

(I) Wald's proof

Let A be the m by n matrix with elements a_{ij}. And we denote by r an m-dimensional column vector with components r_i and by s an n-dimensional column vector with components s_j. p and q represent the column vectors with components p_j $(j = 1, 2, \cdots, n)$ and q_i $(i = 1, 2, \cdots, m)$ respectively.

Wald considers the following system, which we denote by (W).

$$\text{(W)} \begin{cases} As \leqq r, \\ {}^tAq = p, \\ p_j = f_j(s_j) & (j = 1, 2 \cdots, n), \\ {}^tq(As - r) = 0. \end{cases}$$

Wald makes the following assumptions.

(A1) $A \geqq 0$, and for each j there is at least one i for which $a_{ij} \neq 0$.

(A2) $r > 0$.

(A3) For each j, the function $f_j(s_j)$ is defined for every positive value of s_j, its value is nonnegative, continuous , and strictly monotone decreasing, and in addition $\lim_{s_j \to 0} f_j(s_j) = +\infty$.

Wald's theorem If assumptions (A1) to (A3) are satisfied, then there exist solutions $s_j^* > 0$, $p_j^* \geqq 0$ $(j = 1, 2, \cdots, n)$, $q_i^* \geqq 0$ $(i = 1, 2, \cdots, m)$ for the system (W).

We state three lemmas before embarking on to the proof of Wald's theorem.

Lemma 1 Let s_j^*, p_j^* $(j = 1, 2, \cdots, n)$, q_i^* $(i = 1, 2, \cdots, m)$ be solutions and let $s_j^* + \Delta s_j$, $p_j^* + \Delta p_j$ $(j = 1, 2, \cdots, n)$, $q_i^* + \Delta q_i$ $(i = 1, 2, \cdots, m)$ be solutions for the system that arise out of (W) if the r_i are replaced by $r_i + \Delta r_i$. Then the following condition holds;

$$\Delta q_1 \Delta r_1 + \Delta q_2 \Delta r_2 + \cdots + \Delta q_m \Delta r_m \leqq 0.$$

Lemma 2 Let $\{r_1^k, r_2^k, \cdots, r_m^k\}$ $(k = 1, 2, \cdots)$ be a convergent sequence of m-tuples of positive values that converges toward the m-tuple $r_1, r_2, \cdots, r_m$, and let s_j^k, p_j^k $(j = 1, 2, \cdots, n)$, q_i^k $(i = 1, 2, \cdots, m)$ be solutions for the system (W^k) which are obtained by substituting r_i^k for r_i in (W). Then the sequence $\{s_j^k\}$ is bounded and for every i the sequence $\{q_i^k\}$ is also bounded.

Lemma 3 If solutions of the system (W^k) converge as $k \longrightarrow +\infty$ to s_j^*, p_j^*, q_i^*, then these numbers constitute solutions to the system (W).

We present a brief sketch of his proof.

Proof. For $n = 1$, the system obviously has a solution. Indeed, we define

$$s_1^* = \max\{\alpha > 0 \mid a_{i1}\alpha \leqq r_i,\ i = 1, 2, \cdots, m\}$$

Then we obtain $p_1^* = f_1(s_1^*) \geqq 0$. We can choose q_i^* $(i = 1, 2, \cdots, m)$ from the non-empty convex set

$$\{(q_1, \cdots, q_m) \mid p_1^* = a_{11}q_1 + \cdots + a_{m1}q_m \text{ and } q_i = 0 \text{ if } r_i - a_{i1}s_1^* > 0\}.$$

We prove Wald's theorem for the case n on the inductive hypothesis that it holds for $n - 1$.

We define $\overline{\lambda} = \max\{\lambda > 0 \mid r_i - a_{in}\lambda \geqq 0,\ \ i = 1, 2, \cdots, m\}$. For each $\lambda \in (0, \overline{\lambda})$, consider the system, which we denote by $(\mathrm{W}_{n-1}^{\lambda})$,

$$(\mathrm{W}_{n-1}^{\lambda}) \begin{cases} a_{i1}s_1 + \cdots + a_{i,n-1}s_{n-1} \leqq r_i - a_{in}\lambda = r_i' \quad (i = 1, 2, \cdots, m), \\ p_j = f_j(s_j), \\ p_j = a_{1j}q_1 + \cdots + a_{mj}q_m \quad (j = 1, 2, \cdots, n-1), \\ q_i(a_{i1}s_1 + \cdots + a_{i,n-1}s_{n-1} - r_i') = 0. \end{cases}$$

For $\lambda \in (0, \overline{\lambda})$, the system $(\mathrm{W}_{n-1}^{\lambda})$ satisfies assumptions (A1) to (A3). Thus the system $(\mathrm{W}_{n-1}^{\lambda})$ has solutions $s_j^* > 0$, $p_j^* \geqq 0$ $(j = 1, 2, \cdots, n-1)$ and $q_i^* \geqq 0$ $(i = 1, 2, \cdots, m)$ by the inductive hypothesis.

Now we define the set for all $\lambda \in (0, \overline{\lambda})$,

$$\Pi(\lambda) = \{a_{1n}q_1 + \cdots + a_{mn}q_m \mid (q_1, \cdots, q_m) \text{ are solutions of } (\mathrm{W}_{n-1}^{\lambda})\}.$$

From lemma 1, it is verified that if $0 < \lambda_1 < \lambda_2 < \overline{\lambda}$, then no number in $\Pi(\lambda_2)$ is less than some number in $\Pi(\lambda_1)$. Indeed, let λ_1 and λ_2 be two numbers such that $0 < \lambda_1 < \lambda_2 < \overline{\lambda}$. If $(q_1, q_2, \cdots, q_m)$ is any solution set of $(\mathrm{W}_{n-1}^{\lambda_1})$ and $(q_1 + \Delta q_1, q_2 + \Delta q_2, \cdots, q_m + \Delta q_m)$ is any solution set of $(\mathrm{W}_{n-1}^{\lambda_2})$, then lemma 1 shows that

$$-a_{1n}(\lambda_2 - \lambda_1)\Delta q_1 - \cdots - a_{mn}(\lambda_2 - \lambda_1)\Delta q_m \leqq 0.$$

Then we obtain

$$a_{1n}\Delta q_1 + a_{2n}\Delta q_2 + \cdots + a_{mn}\Delta q_m \geqq 0.$$

Wald ([53], p.15) shows that the set $\Pi(\lambda)$ is bounded, closed and convex. So the set $\Pi(\lambda)$ is closed interval, which we denote by $[\underline{\pi}(\lambda), \overline{\pi}(\lambda)]$. From this, it follows that the sets $\Pi(\lambda_1)$ and $\Pi(\lambda_2)$ are either disjoint or they have only one point in common.

Now let $\{\lambda_k\}$ $(k = 1, 2, \cdots)$ be a sequence such that for every k, $0 < \lambda < \lambda_k < \overline{\lambda}$ and $\lim_{k \to +\infty} \lambda_k = \lambda$. The sequence $\{\underline{\pi}(\lambda_k)\}$ is bounded. Let β be a limit of this sequence. Then $\beta \geqq \overline{\pi}(\lambda)$. From lemma 2, the sequence of solution sets for the system $(\mathrm{W}_{n-1}^{\lambda_k})$ $(k = 1, 2, \cdots)$ is bounded. From lemma 3, β belongs to the set $\Pi(\lambda)$. Since $\lambda_k > \lambda$, it must be identical with $\overline{\pi}(\lambda)$. Hence we have $\lim_{k \to +\infty} \underline{\pi}(\lambda_k) = \overline{\pi}(\lambda)$. Similarly, from $0 < \lambda_k < \lambda < \overline{\lambda}$ and $\lim_{k \to +\infty} \lambda_k = \lambda$, it follows that $\lim_{k \to +\infty} \overline{\pi}(\lambda_k) = \underline{\pi}(\lambda)$. Furthermore, it follows that when $\lambda \longrightarrow \overline{\lambda}$ the sequence $\{\underline{\pi}(\lambda)\}$ converges.

For every $\lambda \in (0, \overline{\lambda})$, we set

$$\underline{\tau}(\lambda) = \underline{\pi}(\lambda) - f_n(\lambda),$$
$$\overline{\tau}(\lambda) = \overline{\pi}(\lambda) - f_n(\lambda),$$

and define the closed interval $T(\lambda) = [\underline{\tau}, \overline{\tau}]$. Let $0 < \lambda_1 < \lambda_2 < \overline{\lambda}$. Then it follows that $\overline{\tau}(\lambda_1) < \underline{\tau}(\lambda_2)$ by assumption (A3). It therefore follows that $T(\lambda_1)$ and $T(\lambda_2)$ are disjoint for $\lambda_1 \neq \lambda_2$.

We distinguish two cases; one is the case in which $\lim_{\lambda \to \overline{\lambda}} \underline{\pi}(\lambda) > f_n(\overline{\lambda})$ holds, the other is the case in which $\lim_{\lambda \to \overline{\lambda}} \underline{\pi}(\lambda) \leqq f_n(\overline{\lambda})$ holds. We focus on the former case. We omit the latter case because the proof is straightforward and that this does not matter for the purpose of our arguments.

In the former case, it is obvious that $\lim_{\lambda \to \overline{\lambda}} \underline{\tau}(\lambda) > 0$ and $\lim_{\lambda \to 0} \overline{\tau}(\lambda) = -\infty$. Then, it follows that there exists a $\lambda_0 \in (0, \overline{\lambda})$ such that $0 \in T(\lambda_0)$.

Then $s_n = \lambda_0$ together with the solution of ($\mathrm{W}_{n-1}^{\lambda_0}$) constitute the solution of (W). This completes the proof of Wald's theorem.

As we have already said in section 5, the essential argument of Wald's proof is to show that there exists a $\lambda_0 \in (0, \overline{\lambda})$ such that $0 \in T(\lambda_0)$. As Hildenbrand [20] says, Wald applies the intermediate value theorem for a correspondence in order to obtain this result. He presents the theorem without proof. We prove here the intermediate value theorem for a correspondence.

Intermediate value theorem Let T be a convex-valued and bounded-valued correspondence from a closed interval I into $\mathbb{R}$ that has a closed graph. If there are $\underline{x}$ and $\overline{x}$ in I such that $T(\underline{x})$ contains a non-positive number and $T(\overline{x})$ contains a nonnegative number, then there exists $x^* \in I$ such that $0 \in T(x^*)$.

Proof. Let $I = [\underline{x}, \overline{x}]$. It suffices to prove the theorem for the case where $0 \notin T(\underline{x})$. Without loss of generality, we may assume that $T(\underline{x})$ contains only negative numbers and that $T(\overline{x})$ contains only positive numbers.

Now we suppose that there exists no $x^* \in I$ such that $0 \in T(x^*)$.

By assumption, T is a convex-valued correspondence. Then the interval I can be decomposed into the union of some nonintersecting nonempty interval I^- and I^+ of I. Here an interval I^- consists of all points x such that $T(x)$ contains only negative numbers. Similarly, an interval I^+ consists of all points x such that $T(x)$ contains only positive numbers. It is clear that T is upper hemi-continuous since T is a bounded-valued correspondence and has a closed graph. By the upper hemi-continuity of T, both I^- and I^+ are relatively open in I. Moreover, I^+ and I^- are disjoint, i.e. $I^+ \cap I^- = \emptyset$, and $I^+ \cup I^- = I$. This contradicts to the connectedness of I.

(II) The Minkowski-Farkas lemma

Based upon Wald's idea of proof, we can prove the Minkowski-Farkas lemma with the additional conditions (A1) $A \geqq 0$, and for each j there is at least one i for which $a_{ij} \neq 0$, and (A2) $r > 0$.

The Minkowski-Farkas lemma asserts that exactly one of the following alternatives holds

(11) $As = r$

has a nonnegative solution, or the inequalities

(12) ${}^tAq \geqq 0, \ {}^tqr < 0$

have a solution.

Let us consider the column vectors a_i of A. The set of all nonnegative linear combinations of the a_i forms a convex cone $\angle\{a_1, a_2, \cdots, a_n\}$ spanned by the a_i. The statement that the equality (11) has no nonnegative solution means that vector r does not lie in $\angle\{a_1, a_2, \cdots, a_n\}$. In this case, the lemma asserts that there exists a vector q which makes an obtuse angle with r and a nonobtuse angle with each of the vector a_i. This implies that there exists a hyperplane that has the cone $\angle\{a_1, a_2, \cdots, a_n\}$ on one side and the point r on the other. For this reason, the Minkowski-Farkas lemma is referred to as the theorem of the separating hyperplane.

A convex set in question is build up from the convex hull of a finite number of halflines. Therefore we can construct such a hyperplane by finite steps. This fact enabled us to prove the theorem on mathematical induction.

Proof. Assuming now that (11) has no nonnegative solution, we shall show that (12) has a solution. We proceed by induction on n, the number of columns of A.

If $n = 1$, then (11) becomes

$$\begin{cases} a_{11}s_1 = r_1 \\ a_{21}s_1 = r_2 \\ \quad\vdots \qquad \vdots \\ a_{m1}s_1 = r_m. \end{cases}$$

Since a_1 and r are linearly independent, we can choose $q \neq 0$ so as to satisfy (12).

Now, assume the theorem holds for $n = N - 1$ and let us prove it for $n = N$.

If the solution q satisfies

$$a_{1N}q_1 + a_{2N}q_2 + \cdots + a_{mN}q_m \geqq 0,$$

then the assertion is proved.

Conversely, suppose

$$a_{1N}q_1 + a_{2N}q_2 + \cdots + a_{mN}q_m < 0.$$

Let

$$\overline{a_j} = \langle q, a_N\rangle a_j - \langle a_j, q\rangle a_N \quad (1 \leqq j \leqq N-1),$$
$$\overline{r} = \langle q, a_N\rangle r - \langle q, r\rangle a_N.$$

Then $\overline{r} \notin \angle\{\overline{a_1}, \overline{a_2}, \cdots \overline{a_{N-1}}\}$.

For if not so, there exist nonnegative numbers λ_j such that $\overline{r} = \sum_{j=1}^{N-1} \lambda_j \overline{a_j}$. Then

$$\langle q, a_N\rangle r - \langle q, r\rangle a_N$$

$$= \sum_{j=1}^{N-1} \lambda_j [\langle q, a_N\rangle a_j - \langle a_j, q\rangle a_N]$$

$$= \sum_{j=1}^{N-1} (\lambda_j \langle q, a_N\rangle a_j) + \sum_{j=1}^{N-1} (-\lambda_j \langle a_j, q\rangle a_N).$$

Hence we have

$$r = \sum_{j=1}^{N-1} \lambda_j a_j + \frac{\langle q, r\rangle + \sum_{j=1}^{N-1} (-\lambda_j \langle a_j, q\rangle)}{\langle q, a_N\rangle} a_N,$$

which means $r \in \angle\{a_1, a_2, \cdots a_{N-1}, a_N\}$. This is a contradiction.

By the inductive hypothesis there exists $\overline{q}$ such that $\langle \overline{a_j},\ \overline{q}\rangle \geqq 0\ (1 \leqq j \leqq N-1)$ and $\langle \overline{q},\ \overline{r}\rangle < 0$.

Then let $q^* = \langle q, a_N\rangle \overline{q} - \langle \overline{q}, a_N\rangle q$ and note

$$\begin{aligned}\langle q^*, a_j\rangle &= \langle q, a_N\rangle\langle \overline{q}, a_j\rangle - \langle \overline{q}, a_N\rangle\langle q, a_j\rangle \\ &= \langle \overline{a_j}, \overline{q}\rangle \geqq 0,\end{aligned}$$

$$\begin{aligned}\langle q^*, a_N\rangle &= \langle q, a_N\rangle\langle \overline{q}, a_N\rangle - \langle \overline{q}, a_N\rangle\langle q, a_N\rangle \\ &= 0,\end{aligned}$$

$$\begin{aligned}\langle q^*, r\rangle &= \langle q, a_N\rangle\langle \overline{q}, r\rangle - \langle \overline{q}, a_N\rangle\langle q, r\rangle \\ &= \langle \overline{q}, \overline{r}\rangle < 0.\end{aligned}$$

Thus, q^* satisfies (12) and the theorem is proved.

(III) The duality theorem in linear programming

Moreover, as is well known, the standard method to prove the duality theorem in linear programming is to appeal to the Minkowski-Farkas lemma. We replace the equality of the price-cost conditions for goods ${}^tAq = p$ by the inequality ${}^tAq \geqq p$ and p is assumed to be given. We introduce the complementary slackness conditions (16) ${}^ts({}^tAq - p) = 0$. It can be verified that Wald's idea is applicable to the proof of the duality theorem in linear programming.

We consider the following system.

$$\begin{aligned}&(13)\quad As \leqq r,\\ &(14)\quad {}^tq(As - r) = 0,\\ &(15)\quad {}^tAq \geqq p,\\ &(16)\quad {}^ts({}^tAq - p) = 0.\end{aligned}$$

We assume (A1) $A \geqq 0$, and for each j there is at least one i for which $a_{ij} \neq 0$, (A2') $r \geqq 0$ and $p \geqq 0$. We prove that the above system has nonnegative solutions s and q.

Proof. We proceed by induction on n. If $n = 1$, then (13) becomes

$$\begin{cases} a_{11}s_1 \leqq r_1 \\ a_{21}s_1 \leqq r_2 \\ \quad\vdots \qquad \vdots \\ a_{m1}s_1 \leqq r_m. \end{cases}$$

By assumption (A1), for each j there is at least one i such that $a_{ij} \neq 0$. For each such i we form r_i/a_{i1} and denote the minimum of these values by α. Then α is a solution of (13).

For the case $s_1 = \alpha = 0$, it is obvious that we can choose q in a manner satisfying conditions (14)(15) and (16).

We consider the case $s_1 = \alpha > 0$. For $s_1 = \alpha > 0$, if (13) $a_{i1}s_i \leqq r_i$ $(i = 1, 2, \cdots, m)$ hold with strict inequality for some i, then the corresponding q_i is zero. Then, without loss of generality, we may assume (reordering if necessary)

$$q = {}^t(\underbrace{q_k}_{k}, \underbrace{0}_{(m-k)}), \quad q_k \neq 0.$$

In view of (15) and (16), we can choose q_k so as to satisfy the condition $a_{11}q_1 + a_{21}q_2 + \cdots + a_{k1}q_k = p_1$.

Now we define

$$\overline{\lambda} = \max\{\lambda \geqq 0 \mid r_i - a_{iN}\lambda \geqq 0, \ \ i = 1, 2, \cdots, m\}$$

and $r_i' = r_i - a_{iN}\lambda$. From $\lambda \in [0, \overline{\lambda}]$, it follows that $r_i' \geqq 0$.

Consider the system for $n = N - 1$.

$$(17) \begin{cases} a_{11}s_1 + a_{12}s_2 + \cdots + a_{1,N-1}s_{N-1} & \leqq r_1 - a_{1N}\lambda = r_1' \\ a_{21}s_1 + a_{22}s_2 + \cdots + a_{2,N-1}s_{N-1} & \leqq r_2 - a_{2N}\lambda = r_2' \\ \quad \vdots \qquad\qquad \vdots & \ \vdots \\ a_{m1}s_1 + a_{m2}s_2 + \cdots + a_{m,N-1}s_{N-1} & \leqq r_m - a_{mN}\lambda = r_m' \\ & \lambda \in [0, \overline{\lambda}], \end{cases}$$

$$(18) \quad q_1(\sum_{j=1}^{N-1} a_{1j}s_j - r_1') + \cdots + q_m(\sum_{j=1}^{N-1} a_{mj}s_j - r_m') = 0,$$

$$(19) \begin{cases} a_{11}q_1 + a_{21}q_2 + \cdots + a_{m1}q_m & \geqq p_1 \\ a_{12}q_1 + a_{22}q_2 + \cdots + a_{m2}q_m & \geqq p_2 \\ \quad \vdots \qquad\qquad \vdots & \ \vdots \\ a_{1,N-1}q_1 + a_{2,N-1}q_2 + \cdots + a_{m,N-1}q_m & \geqq p_{N-1}, \end{cases}$$

$$(20) \quad s_1(\sum_{i=1}^{m} a_{i1}q_i - p_1) + \cdots + s_{N-1}(\sum_{i=1}^{m} a_{i,N-1}q_i - p_{N-1}) = 0.$$

We assume that the system for $n = N - 1$ has nonnegative solutions $s_1, \cdots, s_{N-1}$ and $q_1, \cdots, q_m$. And we now prove the theorem for $n = N$.

We distinguish two cases.

[Case1] There exists q in the solution set for $n = N - 1$ such that $a_{1N}q_1 + \cdots + a_{mN}q_m \geqq p_N$. In this case, we can choose $s_N = \lambda \in [0, \overline{\lambda}]$ so as to suit our purposes. Then q together with $s_1, \cdots, s_{N-1}, s_N = \lambda$ constitute the solution of the system.

[Case2] For any q in the solution set, $a_{1N}q_1 + \cdots + a_{mN}q_m < p_N$ holds. By assumption (A1), there exists at least one i which attains $\overline{\lambda}$. We choose one of such i and denote it by i^*.

Let $f(t) = a_{1N}q_1 + \cdots + a_{i^*N}(q_{i^*} + t) + \cdots + a_{mN}q_m$, $(t \geqq 0)$. Then, $f(0) < p_N$, $f(+\infty) = +\infty$ and f is continuous. By the intermediate value theorem, there exists t^* such that $f(t^*) = p_N$. Consider $q' = (q_1, \cdots, q_{i^*} + t^*, \cdots, q_m)$. For such a vector q', the condition (18) holds, since $r'_{i^*} = 0$ and $a_{i^*1}s_1 + \cdots + a_{i^*,N-1}s_{N-1} = 0$. It is obvious that q' satisfies the condition (19). The equation (20) also holds. Here we note when $a_{i^*j} = 0$, $s_j \geqq 0$, and when $a_{i^*j} \neq 0$, $s_j = 0$. The left-hand-side of (20) is

$$\begin{aligned} & s_1[a_{11}q_1 + \cdots + a_{i^*1}(q_{i^*} + t^*) + \cdots + a_{m1}q_m - p_1] \\ & + \cdots\cdots + s_{N-1}[a_{1,N-1}q_1 + \cdots + a_{i^*,N-1}(q_{i^*} + t^*) + \cdots + a_{m,N-1}q_m - p_{N-1}] \\ & = 0. \end{aligned}$$

Then q' together with $s_1, \cdots, s_{N-1}, s_N = \overline{\lambda}$ constitute the solution of the system. We complete the proof.

(IV) The minimax theorem

In its mathematical structure, the minimax theorem has turned out to be similar to the duality theorem in linear programming.

We consider the two-person zero-sum matrix game and the minimax theorem. Wald's idea is applicable to the proof of the minimax theorem. We interpret $A \geqq 0$ as a $(m \times n)$ payoff matrix and assume (A1) $A \geqq 0$, and for each j there is at least one i for which $a_{ij} \neq 0$. In order to prove the minimax theorem, it suffices to show that the expected payoff has a saddle point.

Let $\mathbf{1}_m =^t (1, 1, \cdots, 1) \in \mathbb{R}^m$, $\mathbf{1}_n =^t (1, 1, \cdots, 1) \in \mathbb{R}^n$ and let us consider the following system,

$$\begin{aligned} As &\leqq \mathbf{1}_m, \\ {}^t q(As - \mathbf{1}_m) &= 0, \\ {}^t Aq &\geqq \mathbf{1}_n, \\ {}^t s({}^t Aq - \mathbf{1}_n) &= 0. \end{aligned}$$

We can prove that this system has nonnegative solutions s^* and q^* in exactly the same way as the above theorem. For such s^* and q^*, ${}^t s^* \mathbf{1}_n =^t q^* \mathbf{1}_m$ holds. We set $v =^t s^* \mathbf{1}_n =^t q^* \mathbf{1}_m$.

We define

$$\tilde{s}^* = \frac{1}{v} s^*, \quad \tilde{q}^* = \frac{1}{v} q^*.$$

Thus,

$$\tilde{s}^* \in S_1 = \{(\tilde{s}_1, \tilde{s}_2, \cdots, \tilde{s}_n) \in \mathbb{R}^n \mid \sum_{j=1}^{n} \tilde{s}_j = 1,\ \tilde{s}_j \geqq 0\},$$

$$\tilde{q}^* \in S_2 = \{(\tilde{q}_1, \tilde{q}_2, \cdots, \tilde{q}_m) \in \mathbb{R}^m \mid \sum_{i=1}^{m} \tilde{q}_i = 1,\ \tilde{q}_i \geqq 0\}.$$

$$A\tilde{s}^* \leqq \frac{1}{v} \mathbf{1}_m, \quad {}^t A\tilde{q}^* \geqq \frac{1}{v} \mathbf{1}_n.$$

For any $\tilde{s} \in S_1$ and $\tilde{q} \in S_2$,

$$^t\tilde{q}A\tilde{s}^* \leqq {}^t\tilde{q}\frac{1}{v}\mathbf{1}_m = \frac{1}{v},$$

$$^t\tilde{s}A\tilde{q}^* \geqq {}^t\tilde{s}\frac{1}{v}\mathbf{1}_n = \frac{1}{v},$$

Thus, since $^t\tilde{q}^* A\tilde{s}^* = \frac{1}{v}$, we obtain

$$^t\tilde{q}A\tilde{s}^* \leqq {}^t\tilde{q}^* A\tilde{s}^* \leqq {}^t\tilde{q}^* A\tilde{s}.$$

This shows that $(\tilde{s}^*, \tilde{q}^*)$ is a saddle point.

References

1. Alt, F.: Über die Meßbarkeit des Nutzens. Zeitschrift für Nationalökonomie **7**, 161-169 (1936)
2. Alt, F.: Einzelbesprechungen: Otto Kühne, "Exakte Nationalökonomie". Zeitschrift für Nationalökonomie **7**, 689-691 (1936)
3. Amonn, A.: Cassels System der Theoretischen Nationalökonomie. Archiv für Sozialwissenschaft und Sozialpolitik **51**, 322-361 (1924)
4. Arrow, K.J.: von Neumann and the existence theorem for general equilibrium. In: [14], 15-28
5. Arrow, K.J., Debreu, G.: Existence of equilibrium for a competitive economy. Econometrica **22**, 265-90 (1954)
6. Baumol, W.J., Goldfeld, S.M. (eds.): Precursors in mathematical economics. LSE series of reprints of scarce works on Political Economy **19**, London School of Economics, London 1968
7. Bourbaki, N.: Elément d'histoire des mathématiques. 1984. English translation by J.Meldrum: Elements of the history of mathematics. Springer-Verlag 1991
8. Brouwer, L.E.J.: An intuitionist correction of the fixed-point theorem on the sphere. Proceedings of the Royal Society London **213**, 1-2 (1952)
9. Carnap, R.: Philosophical foundations of physics. Basic Books, New York 1966
10. Cassel, G.: Grundriss einer elementaren Preislehre. Zeitschrift für die gesamte Staatswissenschaft **55**, 395-458 (1889)
11. Cassel, G.: Theoretische Sozialökonomie. 1918. English translation: The Theory of Social Economy. Kelly, New York 1967
12. Cornides, T.: Karl Menger's contributions to social thought. Mathematical Social Sciences **6**, 1-11 (1983)
13. Craver, E.: The emigration of the Austrian economists. History of Political Economy **18** (1), 1-32 (1986)
14. Dore, M., Chakravarty, S., Goodwin, R.: John von Neumann and modern economics. Clarendon Press, Oxford 1989
15. Gödel, K.: Über Vollständigkeit und Widerspruchsfreiheit. Ergebnisse eines mathematischen Kolloquiums **3**, 12-13 (1932)

16. Golland, L., Sigmund, K.: Exact thought in a demented time: Karl Menger and his Viennese mathematical colloquium. The Mathematical Intelligencer **22** (1), 34-45 (2000)
17. Hahn, H.: Empiricism, logic, and mathematics. D. Reidel, Dordrecht 1980
18. Hilbert, D.: Grundlagen der Geometrie. 1899. English Translation: The foundation of geometry. Open Court, Illinois 1971
19. Hilbert, D.: Mathematische Problem. Archiv für Math. und Phys. **3**(1) (1901) In: Gesammelte Abhanglungen **3**, Springer-Verlag, 290-329(1970).
20. Hildenbrand, W.: An exposition of Wald's existence proof. In: Ergebnisse eines mathematischen Kolloquiums (E. Dierker, K. Sigmund eds.).51-61, Springer-Verlag 1998
21. Ingrao, B., Israel, G.: The invisible hands. The MIT Press, Cambridge 1990
22. Kant, I.: Kritik der reinen Vernunft. 1781. English translation by P. Guyer and Allen W. Wood: The critique of pure reason. Cambridge Univ. Press, Cambridge 1998
23. Kakutani, S.: A generalization of Brouwer's fixed point theorem. Duke Mathematical Journal **8**, 457-459 (1941)
24. Kuhn, H.W.: On a theorem of Wald. In: Linear inequalities and related systems (H.W. Kuhn, A.W. Tucker eds.). Annals of Mathematical Studies **38**, 265-273 Princeton Univ. Press, Princeton 1956
25. Leonard, R.J.: From parlor games to social sciences. Journal of Economic Literature **23**, 730-761 (1998)
26. Leonard, R.J.: Ethics and the excluded middle: Karl Menger and social sciences in interwar Vienna. Isis **89**, 1-26 (1998)
27. Mayer, H.: Der Erkenntniswert der funktionellen Preistheorien. In: Wirtschaftstheorie der Gegenwart **2**, 147-239 (1932)
28. Menger, K.: Bemerkungen zu den Extragsgesetzen. Zeitschrift für Nationalökonomie **7**, 25-26 (1936)
29. Menger, K.: Weitere Bemerkungen zu den Extragsgesetzen. Zeitschrift für Nationalökonomie **7**, 388-397 (1936)
30. Menger, K.: The formative years of Abraham Wald and his work in geometry. Annals of Mathematical Statistics **23**, 14-20 (1952)
31. Menger, K.: Austrian marginalism and mathematical economics. In: Carl Menger and the Austrian school of economics (J.R. Hicks, W. Weber eds.), 38-60, Oxford Univ. Press, Oxford 1973
32. Menger, K.: Selected papers in logic and foundations, didactics, economics. Reidel, Dordrecht 1979
33. Menger, K.: Reminiscences of the Vienna circle and the mathematical colloquium. Kluwer, Dordrecht 1994
34. Morgenstern, O.: Mathematical economics. Encyclopaedia of Social Sciences **5**, 364-368 (1932)
35. Morgenstern, O.: Logistik und Sozialwissenschaften. Zeitschrift für Nationalökonomie **7**, 1-24 (1936). English translation by D. Intrator: Logistics and the social sciences. In: Selected economic writings of Oskar Morgenstern (A. Scotter ed.), 389-404 (1976)
36. Neisser, H.: Lohnhöhe und Beschäftigungsrad im Marktgleichgewicht. Weltwirtschaftliches Archiv **36**, 415-455 (1932)
37. Punzo, L.F.: von Neumann and Karl Menger's mathematical colloquium. In: [14], 29-65
38. Punzo, L.F.: The school of mathematical formalism and the Viennese circle of mathematical economics. Journal of the History of Economic Thought **13**, 1-18 (1991)

39. Punzo, L.F.: Karl Menger's contribution to the social sciences. 1994. In: [33], 21-25
40. Reid, C.: Hilbert. Springer-Verlag 1970
41. Scarf, H.: The computation of economic equilibria. Yale Univ. Press, New Haven and London 1973
42. Schams, E.: Die Casselschen Gleichungen und die mathematische Wirtschaftstheorie. Jahrbücher für Nationalökonomie und Statistik **127**, 385-400 (1927)
43. Schlesinger, K.: Über die Produktionsgleichungen der ökonomischen Weltlehre. Ergebnisse eines mathematischen Kolloquiums **6**, 10-11 (1935)
44. Schlesinger, K.: Ein Beitrag zu Mengers „Bemerkungen zu den Ertragsgesetzen“. Zeitshrift für Nationalökonomie **6**, 398-400 (1936)
45. Schumpeter, J.A.: History of economic analysis. Oxford Univ. Press, Oxford 1954
46. Sigmund, K.: A philosopher's mathematicians: Hans Hahn and the Vienna circle. The Mathematical Intelligencer **17** (4), 16-29 (1995)
47. Uzawa, H.: Walras' existence theorem and Brouwer's fixed-point theorem. Economic Studies Quartely **13**, 59-62 (1962)
48. von Mises, L.: Notes and recollection. Libertarian Press, South Holland 1978
49. von Neumann, J.: Zur Hilbertschen Beweistheorie. Mathematische Zeitschrift **26**, 1-46 (1927)
50. von Neumann, J.: Über eines Ökonomisches Gleichungssystem und eine Verallgemeinerung des Brouwerschen Fixpunktsatzs. Ergebnisse eines mathematischen Kolloquiums **8**, 73-83 (1938). English translation by G. Morton(Morgenstern): A model of general equilibrium. Review of Economic Studies **13**, 1-9 (1945)
51. von Stackelberg, H.: Zwei Kritische Bemerkungen zur Preistheorie Gustav Cassels. Zeitschrift für Nationalökonomie **4**, 456-472 (1933)
52. Wald, A.: Über das Hilbertsche Axiomensystem der Geometrie. Ergebnisse eines mathematischen Kolloquiums **3**, 23-4 (1931)
53. Wald, A.: Über die eindeutige positive Losbarkeit der neuen Productionsgleichungen. Ergebnisse eines mathematischen Kolloquiums **6**, 12-18 (1935). English translation: On the unique non-negative solvability of the new production equations (Part 1). In: [6], 281-288
54. Wald, A.: Über die Produktionsgleichungen der ökonomischen Wertlehre (Mitteilung II). Ergebnisse eines mathematischen Kolloquiums **7**, 1-6 (1936). English translation: On the production equations of economic value theory (Part 2) In: [6], 289-293
55. Wald, A.: Über einige Gleichungssysteme der mathematischen Ökonomie. Zeitschrift für Nationalökonomie **7**, 637-670 (1936). English translation by Otto Eckstein: On some systems of equations of mathematical economics. Econometrica **19**, 368-403 (1951)
56. Wolfowitz, J.: Abraham Wald, 1902-1950. Annals of Mathematical Statistics **23**, 1-13 (1952)
57. Weintraub, E.R.: On the existence of a competitive equilibrium: 1930-1954. Journal of Economic Literature **21**, 1-39 (1983)
58. Wicksell, K.: Zur Verteidigung der Grenznutzenlehre. Zeitschrift für die gesamte Staatswissenschaft **56**, 586-591 (1900)
59. Wicksell, K.: Lectures on Political Economy. 1934 Reprint: Kelly, New York 1967
60. Wittgenstein, L.: Tractatus Logico-Philosophicus. Routledge, New York 1922
61. Zeuthen, F.: Das Prinzip der Knappheit, technische Kombination und ökonomische Qualität. Zeitschrift für Nationalökonomie **4**, 1-24 (1933)

Subject Index

Abraham Wald, 167
accretive operator, 23

ball-compact, 23
Brenier's Theorem, 8

conjugate, 10
constrained optimization, 98
convex analysis, 167
cyclical monotonicity, 4

diffeomorphism, 65
dual, 2, 3, 12
duality relations, 12

economic theory of incentives, 1
equimeasurable, 19
existence proof, 167

functional differential inclusion, 23

Gangbo, 8

h-convex, 5–7, 11, 16
$\check{h}$-convex, 10
h-convex potential, 14
h-Fenchel transform, 1, 3
$\check{h}$-Fenchel transform, 3
h-subdifferentiable, 5
h-subdifferential, 5
Hardy-Littlewood inequality, 17
Hilbert's formalism, 167

implementable, 15, 16, 19
implicit function theorem, 65
incentive-compatibility, 15
incentive-compatible, 15, 19
incentive-compatible contracts, 15
inversefunctiontheorem, 65

Jacobian Criterion, 98

Kuhn-Tucker, 98

Lagrange, 98
Levin, 4

mass transportations problems, 1
mathmatical finance, 153
maximal monotone operator, 23
Mc Cann and Gangbo, 4
measure-preserving, 19
measure preserving maps, 3
minimal constraint qualification, 98
Monge-Kantorovich, 2
monotone rearrangement, 17, 19
Monte Carlo method, 153
multifunction, 23

non-C^1 analysis, 65, 98

original control, 23

polar factorization, 4, 14
push-forward, 1

quasi-linear utility function, 15

re-allocation principle, 3, 18
relaxation, 2, 23
relaxed control, 23

Spence and Mirrlees, 2
Spence-Mirrlees condition, 1, 15, 16, 18

tangency-path criterion, 98
theory of incentives, 15
the Vienna circle, 167

Young measure, 23
Young's inequality, 4, 5, 13

Instructions for Authors

A. General

1. Papers submitted for publication will be considered only if they have not been and will not be published elsewhere without permission from the publisher and the Research Center of Mathematical Economics.

2. Every submitted paper will be subject to review. The names of reviewers will not be disclosed to the authors or to anybody not involved in the editorial process.

3. The authors are asked to transfer the copyright to their articles to Springer-Verlag if and when these are accepted for publication.

The copyright covers the exclusive and unlimited rights to reproduce and distribute the article in any form of reproduction. It also covers translation rights for all languages and countries.

4. Manuscript must be written in English. One original and 3 sets of photocopies should be submitted to the following address.

Professor Toru Maruyama
Department of Economics
Keio University
2-15-45 Mita, Minato-ku, Tokyo
108-8345 Japan

5. **Offprints**: Twenty-five (25) **offprints** of each paper will be supplied free of charge. Additional offprints can be ordered at cost price.

B. Preparation of Manuscript

1. Manuscripts must be double-spaced (not 1.5), with wide margins (at least 25 mm), and large type (at least 12 point) on one side of A4 paper. Any manuscript that does not meet these requirements will be returned to the author immediately for retyping.

2. All manuscripts would finally be composed using our Springer LaTeX macro package. The Springer macros use standard LaTeX packages - if you do not have these packages refer to www.dante.de/cgi-bin/ctan-index. If authors make manuscripts by word-processing software other than TeX, please follow our styles as possible. For authors who prepare their manuscripts by TeX, we strongly recommend to visit our homepage, http://www.springer.de/economics/authors/interinf.html, where you can download all the necessary macro packages with instructions for LaTeX2e, LaTeX 2.09 and plain tex. For support, please contact `texhelp@springer.de`.

3. **The title page** must include: title; short (running) title of up to 60/85 characters; first names and surnames of all coauthors with superscript numerals indicating their affiliations; full street addresses of all affiliations; address to which proofs should be sent; fax number; and any footnotes referring to the title (indicated by asterisks*).

4. **Summary/abstract**: Each paper must be preceded by a summary/an abstract, which should not exceed 100 words.

5. **The Journal of Economic Literature index number (JEL classification)** should be indicated and the statement of the **2000 Mathematics Subject Classification (MSC) numbers** is desirable. You can check JEL classification with Internet at `http://ideas.uqam.ca/ideas/data/JEL` as well as 2000 MSC numbers at `http://e-math.ams.org/msc`.

6. **Main text**: All tables and figures must be cited in the text and numbered consecutively with Arabic numerals ac-

cording to the sequence in which they are cited. Please mark the desired position of tables and figures in the left margin of the manuscript.

Do not italicize, underscore, or use boldface for headings and subheadings. Words that are to be italicized should be underscored in pencil.

Abbreviations must be spelled out at first mention in summary/abstract and main text. Abbreviations should also be spelled out at first mention in figure legends and table footnotes.

Short equations can be run in with the text. Equations that are displayed on a separate line should be numbered.

7. **References**: The list of references should be in alphabetical order and include the names and initials of all authors (see examples below). Whenever possible, please update all references to papers accepted for publication, preprints or technical reports, giving the exact name of the journal, as well as the volume, first and last page numbers and year, if the article has already been published or accepted for publication.

When styling the references, the following examples should be observed:

Journal article:
1. or [F-M] Freed, D.S., Melrose, R.B.: A mod *k* index theorem. Invent. Math. **107**, 283-299 (1992)

Complete book:
2. or [C-S] Conway, J.H., Sloane, N.J.: Sphere packings, lattices, and groups (Grundlehren Math. Wiss. Bd. 290) Berlin Heidelberg New York: Springer 1988

Single contribution in a book:
3. or [B] Border, K.C.: Functional analytic tools for expected utility theory. In: Aliprantis, C.D. et al. (eds.): Positive operators, Riesz spaces and economics. Berlin Heidelberg New York: Springer 1991, pp. 69-88

8. **Citations in the text** should be either by numbers in square brackets, e.g. [1], referring to an alphabetically ordered and numbered list, or by the author's initials in square brackets, e.g. [F-M], or by author and year in parentheses, e.g. Freed and Melrose (1992). Any of these styles is acceptable if used consistently throughout the paper. In the third system, if a work with more than two authors is cited, only the first author's name plus "et al." need be given and if there is more than one reference by the same author or team of authors in the same year, then a, b, c should be added after the year both in the text and in the list of references.

9. **Tables** are to be numbered separately from the illustrations. Each table should have a brief and informative title. All abbreviations used in a table must be defined in a table footnote on first use, even if already defined in the text. In subsequent tables abbreviations need not be redefined. Individual footnotes should be indicated by superscript lowercase a, b, c, etc. Permission forms must be provided for any tables from previously published sources (same procedure as with figures; see below).

10. **Figures**: If you have access to suitable software, you can design your figures directly on a computer, but creation by other means is of course also possible. In any case the originals should be pasted into the camera- ready copy at the appropriate places and with the correct orientation. If necessary, the figures may be drawn overscale, but in this case suitably reduced copies should be pasted into the script.

If a figure has been published previously, acknowledge its source and submit written permission signed by author and publisher. The source must be included in the reference list. If a permission fee is required, it must be paid by the author. Responsibility for meeting this requirement lies entirely with the author.

The manufacturer's authorised representative in the EU is Springer Nature Customer Service Centre GmbH, Europaplatz 3, 69115 Heidelberg, Germany. If you have any concerns regarding our products, please contact ProductSafety@springernature.com

Printed and bound by CPI Group (UK) Ltd, Croydon, CR0 4YY
25/09/2026
02222857-0005